Geographic Information Systems and Cartographic Modeling

Geographic Information Systems and Cartographic Modeling

C. DANA TOMLIN

School of Natural Resources
The Ohio State University

Prentice Hall, Englewood Cliffs, N.J. 07632

Library of Congress Cataloging-in-Publication Data

Tomlin, C. Dana.
  Geographic information systems and cartographic modeling / C. Dana
Tomlin.
     p.   cm.
  Includes bibliographical references.
  ISBN 0-13-350927-3
  1. Geography—Data processing.   2. Cartography—Data processing.
  G70.2.T64   1990
  910'.28'5—dc20   89-49333
                                                                        CIP

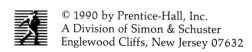

© 1990 by Prentice-Hall, Inc.
A Division of Simon & Schuster
Englewood Cliffs, New Jersey 07632

Printed in the United States of America
10 9 8 7 6 5 4 3

ISBN 0-13-350927-3

Prentice-Hall International (UK) Limited, *London*
Prentice-Hall of Australia Pty, Limited, *Sidney*
Prentice-Hall Canada Inc., *Toronto*
Prentice-Hall Hispanoamericana, S.A., *Mexico*
Prentice-Hall of India Private Limited, *New Delhi*
Prentice-Hall of Japan, Inc., *Tokyo*
Simon & Schuster Asia Pte. Ltd., *Singapore*
Editora Prentice-Hall do Brasil, Ltda., *Rio de Janeiro*

In memory of
a teacher,
a builder,
and a friend

CONTENTS

PREFACE

A *geographic information system* is a facility for preparing, presenting, and interpreting facts that pertain to the surface of the earth. This is a broad definition that can be applied to things ranging from hand-drawn maps to computer systems to groups of people and even to animals. A considerably narrower definition, however, is more often employed. In common parlance, a geographic information system or *GIS* is a configuration of computer hardware and software specifically designed for the acquisition, maintenance, and use of cartographic data.

Cartographic modeling is one approach to the use of this technology. It is a general but well-defined methodology that can be used to address diverse applications in a clear and consistent manner. As the term suggests, cartographic modeling involves *models* (or representations) expressed in *cartographic* form (or as maps). As the term also suggests, cartographic model-*ing* is oriented more toward process than product. Its major concern is not the way in which data are gathered, maintained, or conveyed but the way in which data are used.

This text offers an introduction to cartographic modeling. It is not a general treatise on the broad field of geographic data processing but a generalized treatment of one geographic data-processing methodology. The book is intended for those who already have (or are gaining) experience in routine data storage and retrieval with a geographic information system. In particular, it is intended for those with an interest in using the data-transforming capabilities of this system to answer questions, support decisions, or otherwise solve problems.

The text develops and applies what amounts to a high-level computational language. As a *language*, this is a formal system of symbols, rules governing the formation and transformation of those symbols, and definitions regarding the objects or phenomena they represent. As a *computational* language, its purpose is to represent sequences of automated data-processing activities. And as a *high-level* computational language, it is expressed in terms that transcend the specifics of particular computing systems.

The first three chapters of the text define the structure of this language by establishing conventions. Chapter 1 presents conventions relating to cartographic data, while Chapter 2 describes the manner in which these data are processed. Chapter 3 then indicates how this processing is controlled. In each case, the conventions presented are intended to relate to as many computing systems as possible.

The next three chapters add substance to this structure by introducing capabilities. These are presented in the form of individual operations. Chapter 4 describes operations that transform data as a function of characteristics associated with individual locations. Chapter 5 then presents operations involving functions of neighborhoods, and Chapter 6 covers operations on arbitrary zones. A reference guide to these operations is presented as an appendix.

The final two chapters show how the language can be put to productive use. Here, capabilities are brought together in the form of applied techniques. Chapter 7 presents techniques whose purpose is to describe, while Chapter 8 introduces those whose purpose is to prescribe.

Each of the chapters concludes with a set of exploratory questions. These are intended not so much for exercise as insight. For those whose insight might come from looking back, from peering ahead, or from glancing left and right, a bibliographic section is also included.

Cartographic modeling is a game of only several pieces and a few basic rules but unlimited possibilities. It is also a game that generally requires no previous experience in computer programming, advanced mathematics, or even formal cartography. What the game does require, however, is an eye for both spatial and logical structure.

These pages are the result of a long-held interest that has always enjoyed much-appreciated support. The idea of cartographic modeling was initially inspired by the work and encouragement of Professors David F. Sinton and Carl F. Steinitz at The Harvard Graduate School of Design. Its first exposure to the general public was at the urging of Professor Joseph K. Berry at The Yale School of Forestry and Environmental Studies. And its ongoing development has been supported not only by both of these institutions but also by The Harvard Forest, The University of Connecticut, and The Ohio State University.

To all of those who have been involved, one can only offer thanks. To those about to become involved, one can also offer a challenge. This is a field whose basic foundations are just now beginning to settle, a field whose full potential has certainly not yet been explored. While exploration has its risks, it also has its rewards, the most precious of which can sometimes come from the search as much as the find.

Petersham, Massachusetts   C.D.T.

INTRODUCTION

*Now take the map of surface water. Place a piece of tracing
paper over it and draw rings around all lakes and streams at a
distance of 200 meters. If you will then shade the area within
those rings, this tracing can be combined with the tracings of
steep slopes, sensitive views, high land costs, and so on. To do
so, align the sheets on top of one another and place the whole
stack on a light table or up against a window. The lightest
areas will be those that can best accommodate the proposed
land development.*

- Anon.

This sort of advice can be found in any of a number of publications on environmental planning. It refers to a process in which geographic data
are organized and manipulated in the form of single-factor maps called
overlays. In the typical case, a data base will initially include overlays of factors such as topographic conditions, urban structures,
demographic characteristics, and so on. These will have been derived
either from direct observation or from sources such as published maps or
aerial photographs, and they will all be registered with respect to a
common cartographic base. In this form, existing overlays can be selec-
tively retrieved and graphically transformed to generate new overlays
of site characteristics. The process is called *overlay mapping*.

Overlay mapping methods have been employed since early in
the twentieth century. Only in the 1960s, however, did they begin to
gain wide acceptance. Since then, much of the interest in these methods
has been associated with the evolution of geographic information systems. The advent of this technology has presented an opportunity to
substantially refine and extend traditional overlay mapping methods.

Geographic data processing is a field that has grown from roots
in geography, computing, and application areas ranging from the
natural and social sciences to urban planning and environmental
management. It is a field that has grown steadily since the 1960s and

one that continues to grow in terms of the number of practitioners involved, the range of applications addressed, and the sophistication of tasks performed. It is also a field that has now grown to a point where fascination with tools has matured into concern for the way in which these tools are used.

The cartographic modeling approach attempts to generalize and to standardize the use of geographic information systems. It does so by decomposing data-processing tasks into elementary components, components that can then be recombined with ease and flexibility.

To illustrate the nature of this approach, consider a simple (and contrived) example. Suppose we are interested in mapping levels of attraction to a local swimming hole. Suppose, furthermore, that this swimming hole is one of the several ponds depicted in Fig. I-1. If all of these ponds are similar in terms of intrinsic appeal, then distinctions among them for potential swimmers will be a matter of accessibility. In Fig. I-2 is a cartographic image depicting zones of uniform proximity to one particular pond. Fig. I-3 presents a similar image of distance to the nearest of all other ponds. One way to express the relative attraction to that first pond would be to indicate how its distance from any nearby location compares to that of the closest swimming alternative. This is the characteristic mapped in Fig. I-4. The process involved can be expressed in cartographic modeling notation as follows:

ThisPond   = *LocalRating of WhichPond with -0 for 0 2 with 0 for 1*
ThisFar   = *FocalProximity of ThisPond at ...*
ThosePonds   = *LocalRating of WhichPond with -0 for 0 1 with 0 for 2*
ThatFar   = *FocalProximity of ThosePonds at ...*
ThisMuchMore   = *LocalDifference of ThisFar and ThatFar*
ThisMuchMore   = *LocalRating of WhichPond*
                     with ThisMuchMore for 0 with -0 for 1 2

The concepts and methods underlying this process are examined in the following chapters. In cartographic modeling as in any game, one must begin by becoming familiar with the necessary equipment. Next, the basic rules of play must come to be understood. Only then can playing skills and strategies be developed, often through years of practice, observation, and lessons drawn from the past. It is in recognition of this that the text is organized in three parts, respectively covering
- fundamental conventions,
- elementary capabilities, and
- illustrative techniques.

This structure attempts to facilitate both reference and instruction. As such, it works best when familiar enough to tolerate deviation.

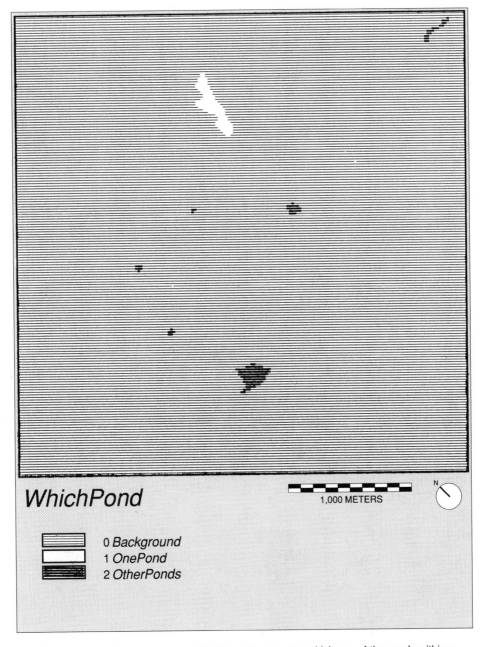

Figure I-1 A map of ponds. *WhichPond* is a map on which one of the ponds within a geographic area is distinguished from its surroundings and from six other ponds by name, number, and graphic symbolism. Figs. I-2 through I-4 illustrate a cartographic modeling process that characterizes the appeal of this pond for use as a swimming hole.

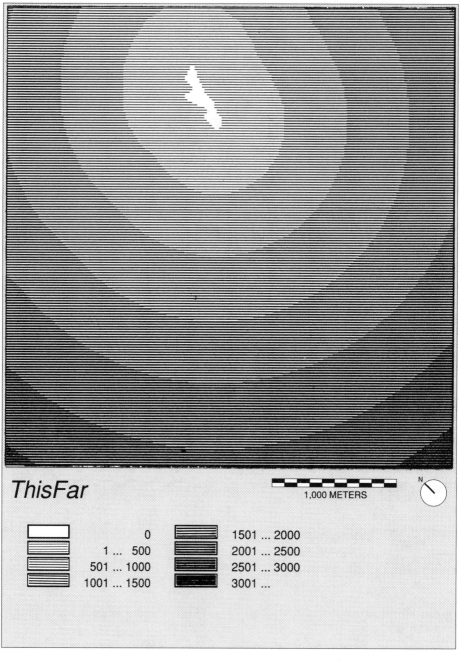

ThisFar

1,000 METERS

N

☐ 0	▤ 1501 ... 2000
▤ 1 ... 500	▤ 2001 ... 2500
▤ 501 ... 1000	▤ 2501 ... 3000
▤ 1001 ... 1500	■ 3001 ...

Figure I-2 A map of proximity to a particular pond. *ThisFar* is a map on which shading patterns represent ranges of distance in meters to one of the the ponds shown in Fig. I-1.

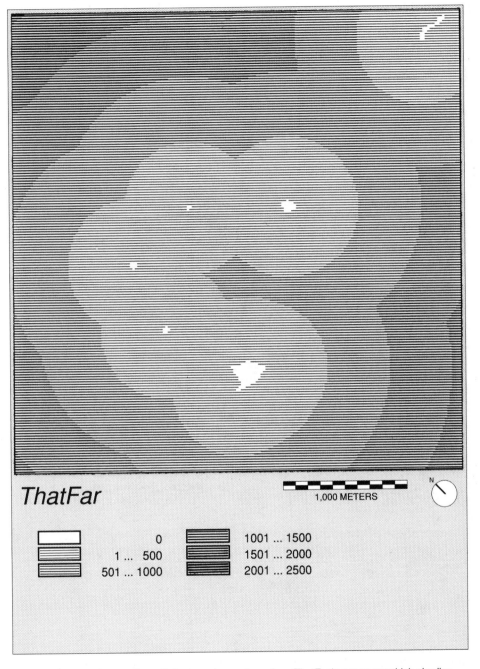

Figure I-3 A map of proximity to a group of ponds. *ThatFar* is a map on which shading patterns represent ranges of distance in meters to the group of six ponds shown in Fig. I-1.

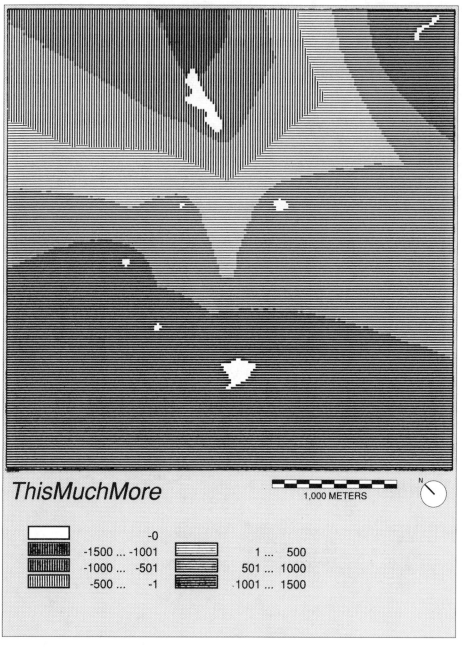

Figure I-4 A map of the extra distance required to access a particular pond. *ThisMuch-More* is a map on which shading patterns represent ranges of difference between each location's proximity to a particular pond and its proximity to other ponds in meters.

Part I

CARTOGRAPHIC MODELING CONVENTIONS

Chapter 1

DATA

Cartographic modeling methods can be implemented (with varying degrees of success) on any of a number of geographic information systems. To do so, however, does require the adoption of certain conventions. The conventions presented here are not those of any particular system. On the contrary, they are intended to relate to as wide a range of computing environments as possible. These conventions are introduced in three chapters, respectively dealing with
- data,
- data processing, and
- data-processing control.

Data are simply recorded facts. In the case of *geographic* data, these are facts pertaining (or like those pertaining) to locations on or near the surface of the earth.

There are a number of ways in which geographic data can be stored for digital processing. These may vary considerably from one geographic information system to another and can be quite significant in terms of specific technical considerations. In terms of the more general concepts and methods of cartographic modeling, however, specifics relating to storage structure need not affect fundamental conventions.

In order to relate to as many actual computing systems as possible, we will establish conventions in terms of a highly generalized data construct. This construct does not dictate the specific structures in which data are actually stored but does indicate the general manner in which data are organized (or appear to be organized) from the perspective of a typical user. The construct is intended to provide a common frame of reference that can easily be translated both to and from a variety of storage formats.

A diagram of this data construct is presented in Fig. 1-1. Note that it is organized as a hierarchy in which certain components are comprised of others. Among the major components defined are

- *cartographic models,*
- *map layers,*
- *titles,*
- *resolutions,*
- *orientations,*
- *zones,*
- *labels,*
- *values,*
- *locations,* and
- *coordinates.*

While certain of these components may seem familiar, beware that each is specially defined for use in the present context. Beware, too, that other terms may also have special meaning. Each of these special terms is presented in **bold italics** when first defined.

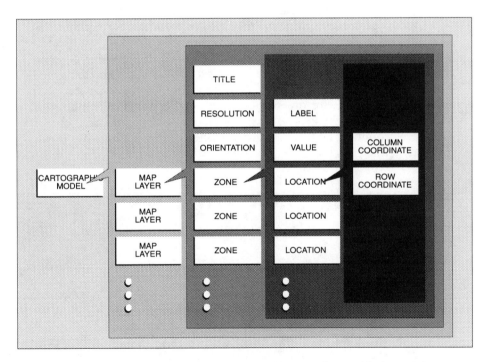

Figure 1-1 A general way of organizing geographic data. Geographic data can be organized as a hierarchy of component parts that correspond to familiar elements in traditional cartography.

1-1 CARTOGRAPHIC MODELS

At the base of this hierarchical tree is the *cartographic model*. A cartographic model can be envisioned as a bound collection of maps much like an atlas. As illustrated in Fig. 1-2, however, it is a collection of maps that are organized such that each of these layers of information pertains to a common site. This site is then referred to as the model's *study area*.

It is important to note that a cartographic model conveys information about its study area in both implicit and explicit form. As we will see, each of the layers of data within a cartographic model will explicitly describe the nature of each location in its study area in terms of a stated characteristic. We will also see that a great deal of additional information that is not explicitly recorded (at least not at first) is nonetheless implicit in the spatial and logical relationships among those data that are recorded and in the meanings attributed to them. This implicit information must also be considered part of a cartographic model. Much of this text will in fact be devoted to the ways in which such information can be converted into explicit form.

The examples and questions that are presented throughout this text all relate to a particular cartographic model. The study area represented by that model is depicted in Fig. 1-3. It is an area of some 13 square kilometers resembling (with a certain amount of cartographic and pedagogic license) an actual site that is located approximately 100 kilometers west of Boston in the village of Petersham (pronounced *Peters-ham*), Massachusetts. The study area takes its name from one of the more prominent bodies of water (and best swimming holes) in town, Brown's Pond.

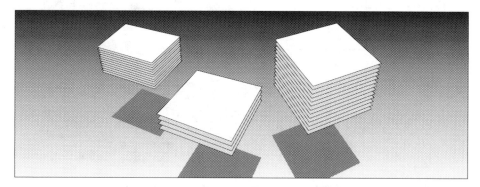

Figure 1-2 A cartographic model. A *cartographic model* is a set of map layers that are all registered with respect to a common cartographic frame of reference.

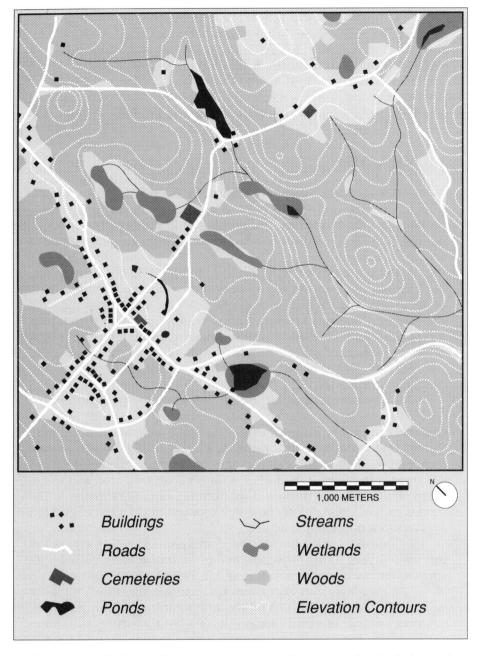

Buildings
Roads
Cemeteries
Ponds

Streams
Wetlands
Woods
Elevation Contours

1,000 METERS

N

Figure 1-3 The Brown's Pond study area. A conventional map depicts the site in central Massachusetts that is used throughout this text to illustrate cartographic modeling conventions, capabilities, and techniques.

Cartographic models may also include data indicating the size, the geodetic position, the history, or other distinguishing characteristics of their study areas. In addition, a model may include data on how, when, or by whom it was created, how it is organized, and so on. While such information may well be important in certain specific contexts, it need not conform to any particular set of conventions. For our purposes, a cartographic model is essentially a set of map layers.

1-2 MAP LAYERS

A *map layer* (or simply a *layer*) is much like a conventional map, a flat drawing indicating the nature, the form, the relative position, and the size of selected conditions in a geographic area. As illustrated in Fig. 1-4, however, a layer is more like a map of just one of an area's characteristics.

On a conventional map like that shown in Fig. 1-3, a given location may well be characterized in terms of two or more attributes or no attributes at all. Each of a map layer's locations, on the other hand, is characterized in terms of exactly one attribute. In these terms, map layers are similar to what are variously called *themes*, *overlays*, *coverages*, *maps*, and *data elements* as well as *layers* in cartography; *images*, *data planes*, or *picture functions* in image processing; and *variables* in statistics.

A typical cartographic model will normally include perhaps several dozen layers. Each will depict the model's study area by characterizing every location within that area in terms of a common theme. One of the layers in a cartographic model may depict a characteristic such as topographic slope, for example, while another describes soil types, a third shows buildings, a fourth gives land cost, and so on. More abstract or interpretive characteristics such as distance to the nearest road or suitability for land development might also be included as layers in a cartographic model.

The model of the Brown's Pond study area initially includes just four map layers. These describe the area's topographic altitude, its surface water, its forest vegetation, and its pattern of development.

A map layer may also carry explanatory information describing its nature, its source, its reliability, and so on. For our purposes, however, the only essential components of a layer are its
- *title*,
- *resolution*,
- *orientation*, and
- *zone(s)*.

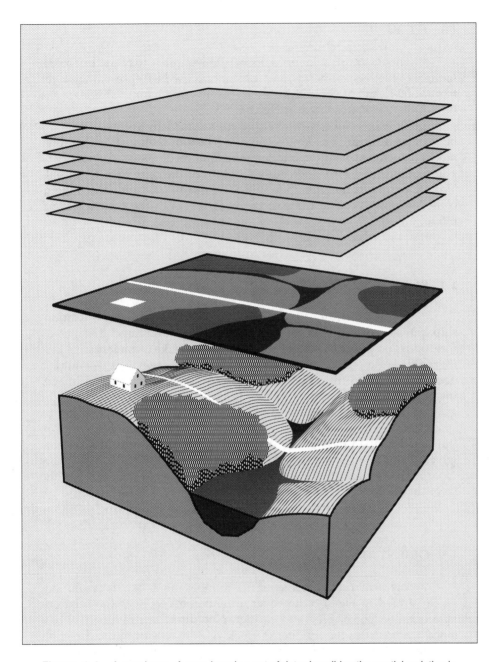

Figure 1-4 A map layer. A *map layer* is a set of data describing the spatial variation in one characteristic of a geographic study area. Here, a layer (center) depicting the major surface conditions within an area is shown in relation to a diagram (below) of the landscape it represents.

1-3 TITLES

The *title* of a layer is its written name. It is an unbroken sequence of letters, numerals, and/or symbols used to represent the layer in various forms of communication. "Landscape," for example, is the title of the layer presented in Fig. 1-5.

Titles are specified when layers are initially created and may be selected arbitrarily as long as no two layers in the same cartographic model have the same title. It is common (and recommended) practice to select titles that clearly indicate the nature of the data they represent. Typical examples include "SoilType," "Version-1.2," and "AverageIncomePerCounty." The titles of the base layers in the Brown's Pond model are *Altitude*, *Water*, *Vegetation*, and *Development*.

1-4 RESOLUTIONS

The *resolution* of a layer is a number describing the relationship between distance as measured "on the ground" and distance as measured "on paper." To that extent, it is much like a conventional cartographic scale. Unlike a conventional scale, however, this number does not relate to the physical size of any particular graphic image. Instead, it specifies the number of feet, meters, or other units of geographic distance that are to be associated with a standard unit of spatial observation. As illustrated in Fig. 1-5, this unit is a *location* (further described in Sec. 1-9), the smallest standard unit of space for which data are recorded.

The resolution of a layer may be set or reset at will. Each of the layers in the Brown's Pond model is at a resolution of 20 meters.

1-5 ORIENTATIONS

The *orientation* of a layer is a number much like its resolution. As illustrated in Fig. 1-5, however, orientation describes the relationship between geographic and cartographic directions rather than distances. It specifies the number of clockwise degrees from the direction depicted as true north to the direction facing the *upper* edge of the cartographic plane (further described in Sec. 1-9).

Like resolution, orientation may be set or reset at will. The Brown's Pond layers are normally oriented at 45 degrees.

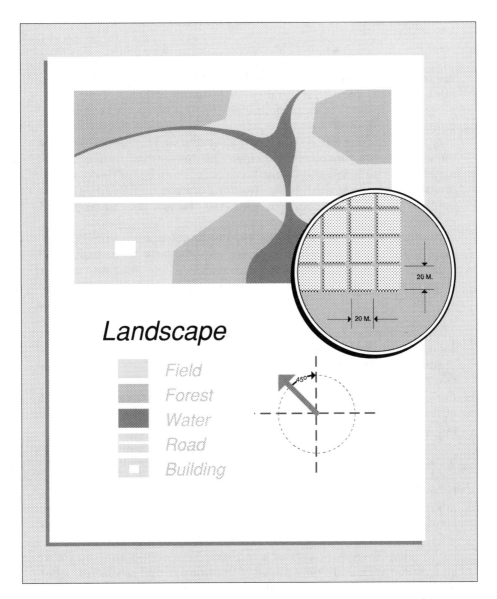

Figure 1-5 A map layer's title, resolution, and orientation. The *title* of a layer is its assigned name. Here, a layer of landscape conditions is entitled *Landscape*. The *resolution* of a layer is a number indicating the relationship between cartographic dimensions and the geographic distances they represent. Here, a resolution of 20 equates the elemental unit of cartographic space with a distance of 20 meters. The *orientation* of a layer is a number indicating the relationship between cartographic directions and the geographic directions they represent. Here, an orientation of 45 indicates that the direction facing the upper edge of the cartographic display is at a clockwise angle of 45 degrees from the direction associated with north.

1-6 ZONES

The final major component of a map layer is a set of one or more *zones*. In common usage, a zone can be defined as a geographic area exhibiting some particular quality that distinguishes it from other geographic areas. The same definition can also be applied to our special use of the term.

Here, a zone is defined as the set of data pertaining to one of the variations on a map layer's theme. The cartographic form of a zone may be large or small and in one piece or in a number of disconnected fragments. As illustrated in Fig. 1-6, for example, the "Building" zone is relatively small and compact, while the "Forest" zone is larger and in several pieces.

Unlike the variations in theme depicted on a conventional map, the zones of a given map layer must be both all-inclusive and mutually exclusive in their spatial coverage. They must account for every part of a study area but must never overlap. In these terms, zones are similar to what are variously called *categories*, *classes*, *data items*, or *regions* in other contexts.

A typical layer will normally include fewer than a dozen zones. This number, however, may vary considerably. A map layer indicating political election results, for example, may have no more than two zones. And a layer of average annual rainfall may have no more than one. A map layer of population density, on the other hand, might well have a separate zone for each of hundreds or even thousands of population density levels.

In the Brown's Pond model, the *Altitude* layer includes zones associated with varying heights above sea level. *Water* includes zones associated with dry land, streams, wetlands, and ponds. The *Vegetation* layer distinguishes nonwooded areas from hardwood, softwood, and mixed-growth forests. And *Development* locates major and minor roads, houses, public buildings, and cemeteries on a background of undeveloped land.

Just like a map layer or a cartographic model, a zone may also be recorded with additional pieces of information that are useful in particular situations but not essential in any general sense. These might include, for example, measurements of the zone's area or perimeter, the position of its centroid, and so on. For our purposes, however, we need only define a zone in terms of three major components. These include the zone's

- *label*,
- *value*, and
- *location(s)*.

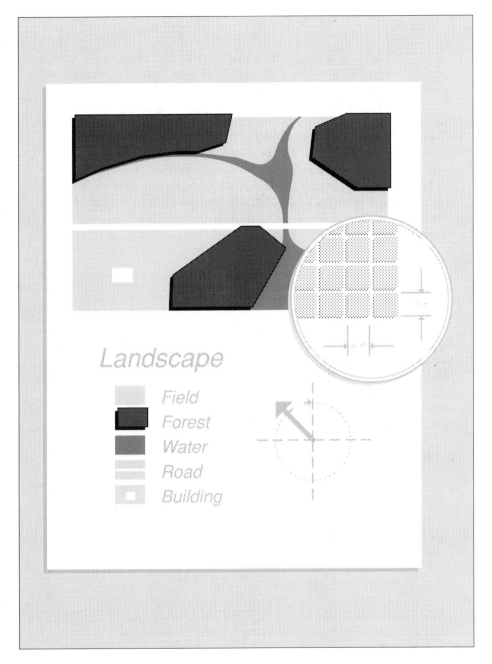

Figure 1-6 A zone. A *zone* is one of the categories into which geographic locations are classified on a map layer. Here, one of the zones (darkened) on a *Landscape* layer is associated with forest vegetation.

1-7 LABELS

The *label* of a zone, much like the title of a layer, is its written name. As illustrated in Fig. 1-7, each label is an unbroken sequence of letters, numerals, and/or symbols.

Labels may be assigned arbitrarily as long as no two zones of the same layer ever share a common label. Each label is normally selected to indicate the nature of its zone. In the Brown's Pond model, for example, the zones of the *Altitude* have labels such as *239-MetersAboveSeaLevel*, *240-MetersAboveSeaLevel*, and so on. The zones of *Water* are labeled *DryLand*, *Streams*, *Wetlands*, and *Ponds*, while those of *Vegetation* are *OpenLand*, *HardWoods*, *SoftWoods*, and *MixedWoods*. And the labels of the zones on the *Development* layer are *VacantLand*, *MajorRoads*, *MinorRoads*, *Houses*, *PublicBuildings*, and *Cemeteries*.

1-8 VALUES

The *value* of a zone is comparable to its label. As illustrated in Fig. 1-7, values represent zones much like colors, symbols, or graphic patterns in traditional cartography. Here, however, the form of representation is neither verbal nor graphic but numeric. With the exception of *null* values (described below), all values are integers. They are similar to what are sometimes called *data codes* or *gray levels* as well as *values* in other computing contexts.

In the Brown's Pond model, zones of the *Altitude* layer are recorded using values of
- 239 for *239-MetersAboveSeaLevel*,
- 240 for *240-MetersAboveSeaLevel*,
- 241 for *241-MetersAboveSeaLevel*,
- 242 for *242-MetersAboveSeaLevel*, and so on up to
- 352 for *352-MetersAboveSeaLevel*;

while those of the *Water* layer are recorded with
- 0 for *DryLand*,
- 1 for *Streams*,
- 2 for *Wetlands*,
- 3 for *Ponds*;

those of the *Vegetation* layer are recorded with
- 0 for *OpenLand*,
- 1 for *HardWoods*,
- 2 for *SoftWoods*,
- 3 for *MixedWoods*;

and those of *Development* are recorded with
- 0 for *VacantLand*,
- 1 for *MajorRoads*,
- 2 for *MinorRoads*,
- 3 for *Houses*,
- 4 for *PublicBuildings*,
- 5 for *Cemeteries*.

Note that values, like labels, can be arbitrarily assigned as long as no two zones of a common layer ever share a common value. Unlike labels, however, values can also be computed. When computation of a value yields anything other than an integer, it is rounded by truncation so that only the integer portion is retained. A ratio of five over three would therefore be calculated not as 1.666 or as two but as one.

Value Precision

The precision with which zonal characteristics can be represented by integer values will depend on the range of values that can be stored and processed by a given computing system. While this will vary from one system to another, we can generally assume that sufficient precision is attainable for most applications.

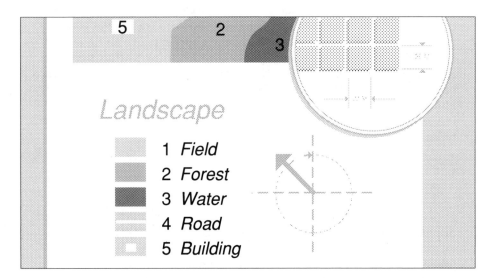

Figure 1-7 A zone's label and value. The *label* of a zone is an assigned name. The *value* of a zone is an assigned or computed number. Here, the values and labels of five *Landscape* zones are darkened.

We can also assume that this range is bounded by known maximum and minimum limits. As a matter of convenience, we will assert that any value above this maximum or below that minimum will automatically be detected and set to the exceeded limit. If the maximum limit were to be 32,767, for example, then the sum of 32,766 plus 234 would not yield 33,000 but 32,767.

Null Value

It is often useful to be able to represent zones of unknown or undefined characteristics in a manner that is qualitatively different from the use of numeric values. We can do this by defining a special *null* value. Unless otherwise specified, null values and the zones they represent are effectively ignored in calculations. Whenever this would make the result of a calculation ambiguous or meaningless, however, that result is also set to the null value. Thus, a null value multiplied by 10 would yield 10, while a null value divided by 10 would be null.

Types of Measurement

Our representation of zones by way of numerical values rather than colors, symbols, or other graphic devices is one of the major distinctions between this data construct and its more traditional counterparts. The use of numbers affords a degree of flexibility and a level of precision that would not be possible otherwise. It also affords the considerable power of mathematical processing. Along with these benefits, however, there also comes a certain responsibility to see that numbers are used in a manner appropriate to that which they represent. It is helpful in this regard to recognize distinctions between four general types of numerical measurement, including

- *ratio*,
- *interval*,
- *ordinal*, and
- *nominal*.

Ratio measurements are those that represent quantities in terms of position relative to a fixed point on a calibrated, linear scale. These measurements are expressed by way of numbers that can generally be transformed or combined with any mathematical function to generate meaningful results. Measurements of characteristics such as age, frequency, physical distances, and monetary value are usually expressed in numbers that relate to a ratio scale.

Consider, however, the measurement of characteristics such as temperature or date of origin. Would it be correct to assert that, in rising from 10 to 20 degrees Centigrade, a temperature has doubled? In degrees Fahrenheit, the same two temperatures would be recorded as 50 and 68. Would it be reasonable to argue that a castle built in the year A.D. 900 is twice as old (or half as young) as a house built in 1800? Suppose we adopt a different calendar.

Both of these examples involve measurements that relate to *interval* scales. They represent quantities in terms of positions that are defined in relation to one another along a calibrated, linear scale but not in relation to any fixed point. As such, these measurements are useful in quantifying differences but not proportions. They are often used to characterize relative (as opposed to absolute) position in space, time, or magnitude in terms of latitudes, longitudes, compass directions, times of day, normalized scores, and so on.

The third type of measurement quantifies differences by order but not by magnitude. Consider, for example, the distinction between first, second, and third place in an athletic competition. These might well be represented by *ordinal* values of one, two, and three. While such values would certainly serve to indicate who beat whom, they would give no indication as to just how much better or worse each competitor actually performed relative to the others or to any absolute standard. The ordinal values represent quantities only in terms of position along an uncalibrated, linear scale. These are often used where quantitative differences are apparent but difficult to measure.

And finally, *nominal* values are those that represent qualities rather than quantities and do so without any reference to a linear scale. Here, numbers are used only as identifiers to distinguish one observation from another. Your telephone number is probably a good example. In representing nominal characteristics, it is customary to use zero for any "background" zone (such as the dry land on a layer of surface water) and consecutive integers beginning with one for all others.

Distinctions among ratio, interval, ordinal, and nominal values become quite significant when these values are to be processed mathematically. Certain functions, such as division, will produce meaningful results only when applied to ratio-scale data. Others, such as subtraction, can safely be applied to interval values as well. Still other functions, such as selecting a maximum or minimum value, can be applied to values of ratio, interval, or ordinal significance. And some mathematical functions, such as testing for equality, can even be applied to nominal data. In general, any function that can be applied to data of a given type in the ratio-interval-ordinal-nominal sequence can also be applied to data of any preceding type in that sequence.

1-9 LOCATIONS

The final major component of a zone is a set of one or more locations. A *location* is the elemental unit of cartographic space for which a label and value are recorded.

A location is, to some extent, like a *cartographic point*: that portion of a plane surface that is uniquely identified by an ordered pair of planar coordinates. If such coordinates are defined with infinite precision, then the planar area associated with a cartographic point can be regarded as immeasurably small and shapeless. If those coordinates are defined over a range of discrete intervals, on the other hand, each point will correspond to a finite portion of the plane. The size and shape of this portion of the plane will depend on the coordinate system involved.

Traditional methods of cartographic representation typically rely on the physical dimensions of a graphic display to define a coordinate system. Since any physical distance can encompass an unlimited number of points, such systems are at least theoretically capable of infinite precision. In actual practice, however, factors such as sheet size, line width, and the limits of human vision impose levels of precision that remain quite finite.

Mathematical representation can also be used to define planar coordinates with unlimited precision by relying on the infinite divisibility of the real number system. Here too, however, such precision can be difficult to achieve in practice. To do so generally requires that the each point's zonal value be expressed as a mathematical function of its two coordinates. While such functions do find considerable use in digital image processing, they are generally quite difficult to establish and to use.

Because of this, most numerical methods of cartographic representation define planar coordinates over a range of discrete intervals that represent finite increments of space. We will adopt the same approach in establishing locations.

For our purposes, all locations are defined with respect to a coordinate system of equal increments along two perpendicular axes. This yields a uniform rectangular grid as illustrated in Fig. 1-8. Note that every location within this grid is associated with one **grid square**. All grid squares are of the same size, shape, and orientation, but each occupies a unique position. The set of all locations of all zones of a given layer will encompass a grid pattern whose overall size and shape will correspond to that of a study area. The cartographic space outside of this study area can then be regarded as an endless zone whose locations conform to the same grid pattern but whose value is null.

In these terms, locations correspond to what are called *pixels* or *picture elements* in image processing, *cells* or *grid cells* in other cartographic contexts, and *observations* in statistics.

Figs. 1-9 through 1-12 depict the pattern of locations associated with each zone of the base layers in the Brown's Pond model.

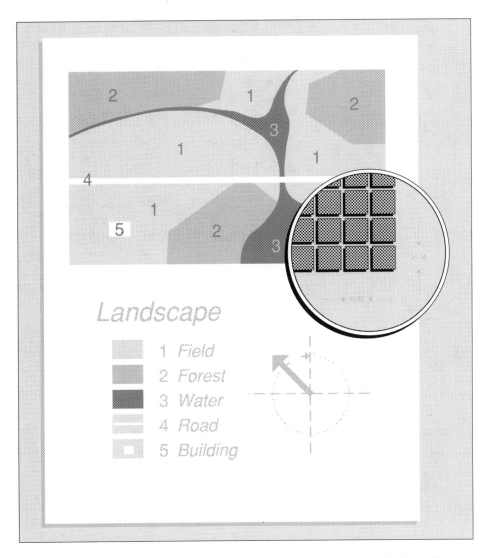

Figure 1-8   Locations. Each of the *locations* on a layer is associated with a unique square unit of cartographic area referred to as its *grid square*. Points halfway between adjacent grid squares are treated as part of the location immediately below and/or to the left.

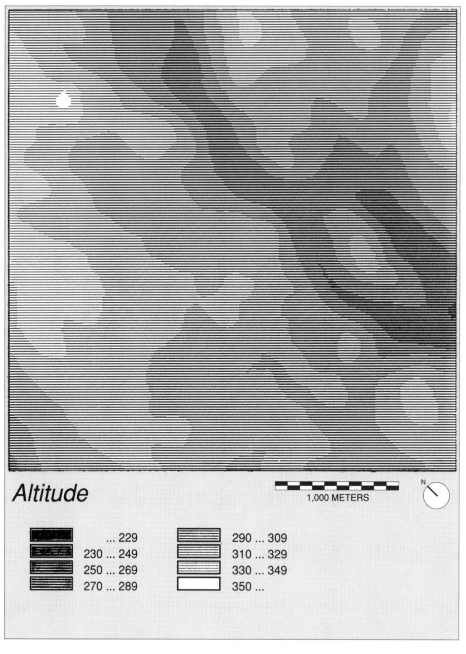

Figure 1-9 A map layer of topographic elevations. *Altitude* is a layer indicating earth surface height above sea level in meters for each location within the Brown's Pond study area. Note that each shade of gray represents not just one zone but a range of elevations.

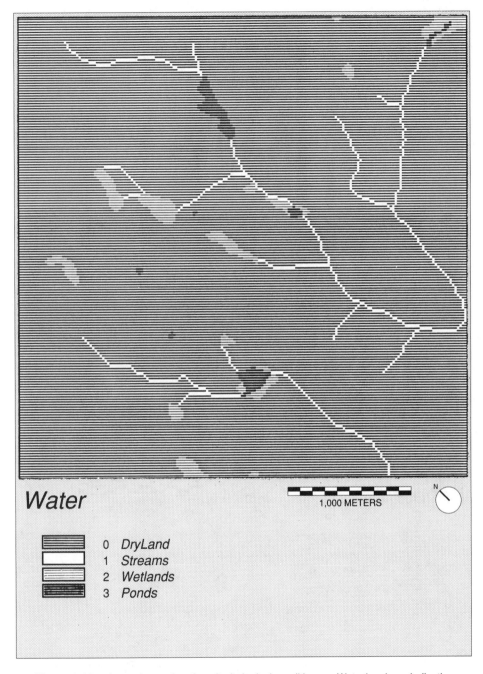

Figure 1-10   A map layer of surface hydrological conditions.   *Water* is a layer indicating the type of surface water, if any, at each location within the Brown's Pond study area.

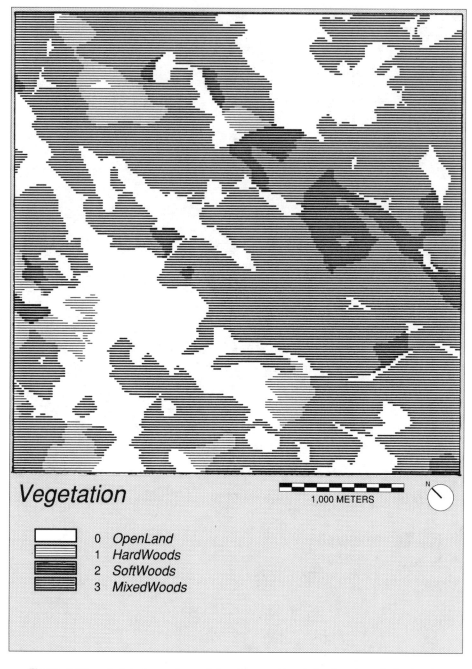

Figure 1-11 A map layer of vegetation types. *Vegetation* is a layer indicating the predominant type of tree cover, if any, at each location within the Brown's Pond study area.

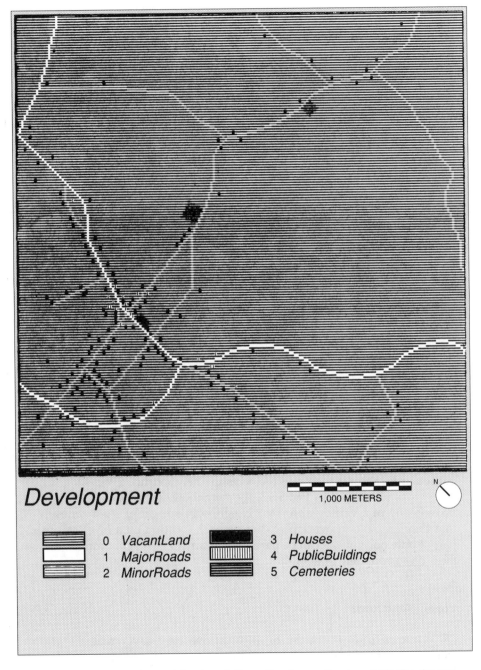

Development

1,000 METERS

N

	0 VacantLand		3 Houses
	1 MajorRoads		4 PublicBuildings
	2 MinorRoads		5 Cemeteries

Figure 1-12 A map layer of developed sites. *Development* is a layer indicating the type of human artifact, if any, at each location within the Brown's Pond study area.

It is useful to assume that the grid pattern associated with the layers of a given cartographic model will normally be displayed such that one edge of that pattern can consistently be designated as its top. In this way, we can use the terms *upper*, *lower*, *left*, and *right* in a conventional graphic sense to indicate cartographic directions without having to specify the actual geographic directions they represent. We will also use the term *diagonal* to refer to the upper left, upper right, lower left, and lower right directions collectively and *lateral* to collectively refer to directions up, down, left, and right. We will reserve the (yet to be described) terms *horizontal* and *vertical*, however, to distinguish a third dimension perpendicular to the cartographic plane.

As spatial entities, the locations on a layer are related to one another not only in terms of their affiliation with related zones but also in terms of their position. Unlike zonal associations, these positional relationships are seldom explicitly recorded. They are implicitly defined by the relative ordering of locations in the cartographic grid.

To deal with the variety of positional relationships that can exist between locations, it is useful to recognize several generic types of locational groupings. Among the most significant of these are
- *neighborhoods,*
- *columns,* and
- *rows.*

Neighborhoods

A *neighborhood* is a set of locations that are at specified cartographic distances and/or directions from a particular location. This particular location is then referred to as the neighborhood *focus.*

As illustrated in Fig. 1-13, the neighborhood of a location may take on any of a variety of forms. Neighborhoods differ from zones in that they are able to overlap. Whereas each location on a layer can be part of only one zone, it can be part of any number of neighborhoods that focus on surrounding locations.

Columns and Rows

A *column* is the set of all locations on a layer that are at a common distance from the layer's leftmost or rightmost edge. Similarly, a *row* includes all locations at a common distance from a layer's upper or lower edge.

Figure 1-13 Neighborhoods. A *neighborhood* is a set of locations (dark gray), each of which bears a specified distance and/or directional relationship to a particular location called the neighborhood *focus* (black). This relationship may be specified in terms of maximum distance (upper left), a range of distances (upper right), ranges of directions (lower left), combinations of distance and direction (lower right), and so on.

Relationships Between Cartographic and Geographic Space

By adopting these conventions associated with locations, we can establish an effective means of describing cartographic space. But what of the actual geographic area it represents? Any relationship between cartographic and geographic geometry must involve some sort of projection. In the case of locations, this projection can be described in terms of the shape, the size (resolution), and the orientation of the geographic area represented by a grid square.

For our purposes, we can assume that the cartographic shape of each location's grid square corresponds directly to the geographic shape of the area it represents. This assumption should not be difficult to accept, particularly if the area being mapped is relatively small. Even if the area is hilly, it can be partitioned into square units by the kind of projection shown in Fig. 1-14.

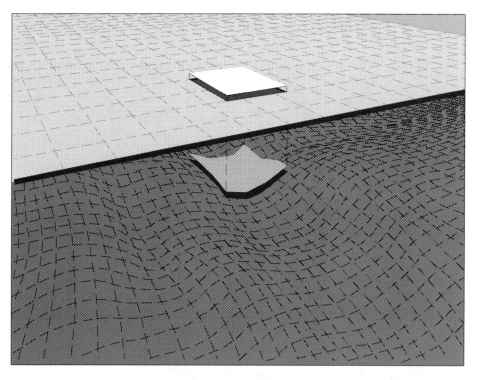

Figure 1-14 The geographic area corresponding to a grid square. By assuming that geographic space (below) can be projected onto a cartographic plane (above) as shown, we can conveniently relate the shape, the orientation, and the size of each location's grid square (white) to similar characteristics of the geographic area it represents (light gray).

As the area being mapped becomes large enough to encompass significant curvature of the earth, however, this assumption that "a square is a square" becomes more troublesome. It simply is not possible to project a round earth onto a flat map without some geometric distortion. The problem is by no means unique to cartographic modeling or to digital mapping. It has, in fact, challenged cartographers for centuries. It has also resulted in an impressive array of sophisticated cartographic projection systems. Each of these systems compromises certain geometric qualities in order to preserve others. By presuming a relationship like that shown in Fig. 1-14, we conveniently circumvent an issue that lies beyond the scope of this text.

The relationship shown in Fig. 1-14 associates each location of a layer with a unique geographic area whose position and spatial form are represented by way of a grid square. This does not necessarily imply, however, that all geographic characteristics represented by locations conform to precise grid boundaries. Seldom, in fact, will that ever be the case. More typically, the use of a location to represent part of a zone indicates only that a cartographic depiction of the zone occupies some portion of that location's grid square.

In order to provide a consistent basis for measurements relating to the position and form of geographic characteristics that are represented by locations, it is useful to make certain inferences regarding spatial configurations *within* grid squares. For any given location, these inferences can often be drawn from the pattern of values associated with adjacent locations. The inferences will vary according to the

- *punctual,*
- *lineal,*
- *areal,* or
- *surficial*

nature of the characteristic involved.

Representing Punctual Characteristics

A *punctual* characteristic is one that can be regarded, for practical purposes, as having no geographic dimensions. It is a characteristic such as the presence of a flagpole, a surveyor's benchmark, or the precise summit of a mountain.

When a location is used to represent a punctual condition, it generally indicates only that the condition occurs somewhere within the location's grid square. Measurements relating to the condition must therefore rely on certain assumptions.

Though punctual characteristics are generally not measurable in terms of spatial form, they can be described in terms of cartographic position. This usually involves measures of distance and/or direction relative to one another or to an external frame of reference. If the cartographic distance between two locations were to be measured from the edges of their respective grid squares, then the distance between two laterally adjacent locations would be zero. This presents a problem in that the distance between two diagonally adjacent locations (or the distance between a location and itself) would also be measured as zero. If the directional relationship between two locations had to be measured from the edges of grid squares, this too would lead to ambiguity.

To resolve these problems in making positional measurements pertaining to the punctual characteristic(s) represented by a location, it is convenient and customary to relate such measurements to the center of that location's grid square. In this way, distances and directions can be measured consistently as shown in Fig. 1-15. Distance and direction are

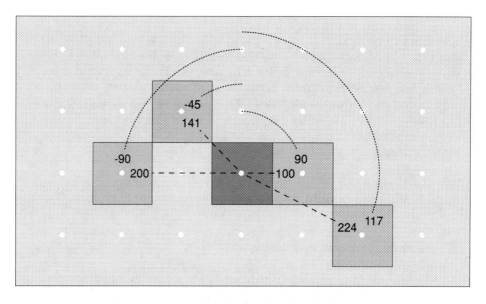

Figure 1-15 Inferring distance and directional relationships between punctual conditions represented by locations. Positional relationships between punctual conditions that are represented by locations can be defined unambiguously by relating distance and/or directional measurements to the centers (white dots) rather than the edges (dark lines) of grid squares. Here, selected distances (dashed lines) and directions (dotted lines) from one location (dark gray) to several neighboring locations (light gray) are expressed in units corresponding to a layer resolution of 100. Directions are expressed in clockwise degrees from an upward bearing.

both calculated by *Euclidean* geometry with respect to the cartographic plane. The Euclidean distance between locations that are located C columns and R rows away from one another on a map layer of resolution X can be expressed as

$$X * (C^2 + R^2)^{1/2}$$

while their directional relationship can be expressed as

$$\arctan (C / R)$$

Representing Lineal Characteristics

Whereas a punctual characteristic has no dimensions, a *lineal* characteristic has one. It is a characteristic such as the presence of a pipeline, a property boundary, or a stream network whose width is small enough to be regarded as negligible and whose fundamental dimension is length.

When a lineal characteristic is represented by a location, it can generally be inferred only that some part of that characteristic occurs within the location's grid square. It is therefore once again useful to relate positional measurements to the centers of grid squares. In this way, distances and directions to surrounding positions can be measured as they would be for punctual characteristics. The only additional convention required here is an understanding that distances and directions between zones are always measured from nearest locations.

The spatial form of a lineal characteristic can be expressed in terms of two major properties: its shape and its size. To infer these properties for a lineal condition represented by a sequence of locations, it is again useful to regard the form of each location as something other than a grid square. One way to do this is to equate consecutive locations with a set of straight line segments between the centers of their adjacent grid squares. In this way, measures of both shape and size can be expressed in terms of discrete increments associated with individual locations. These increments can then be aggregated to characterize overall lineal form.

When the form of a lineal condition is inferred in this manner, the lineal shape represented by any given location can be fully described in terms of the presence or absence of line segments connecting that location to each of from one to eight adjacent neighbors. This yields 2^8 or 256 possible shapes. It can also yield ambiguity, however, whenever three or four of the locations involved are all adjacent. One way to avoid this ambiguity is illustrated in Fig. 1-16. Note here that direct connections between diagonal neighbors are inferred only when such locations are not also connected through a common lateral neighbor. This reduces the number of different line segment configurations that can occur within a single grid square from 256 to 47. Those 47 are illustrated in Fig. 1-17.

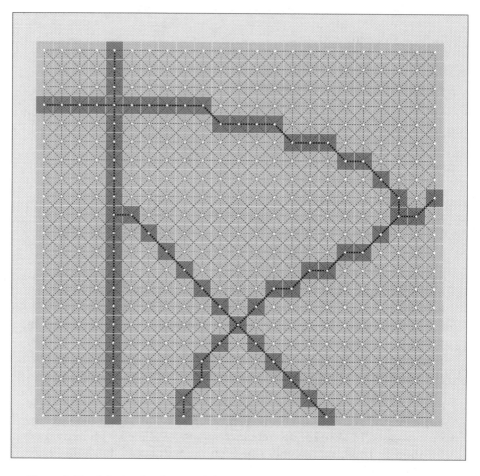

Figure 1-16 Inferring the shape and size of lineal conditions represented by locations. One way to infer the spatial form of a lineal condition represented by locations is to equate the cartographic grid with a network of line segments (dotted lines) connecting the center of each grid square (white dot) to the centers of its adjacent neighbors. In this way, a lineal form represented by a string of locations (darkened) can be equated with the set of line segments (also darkened) that connect its adjacent grid square centers. To avoid ambiguity, this is done such that no connection is inferred between diagonal neighbors when such locations are already linked through a common lateral neighbor.

The most common measure of the size of a lineal condition is its length. One way to estimate the total length of a lineal condition represented by locations is to measure and then add the increments of length that are associated with individual locations.

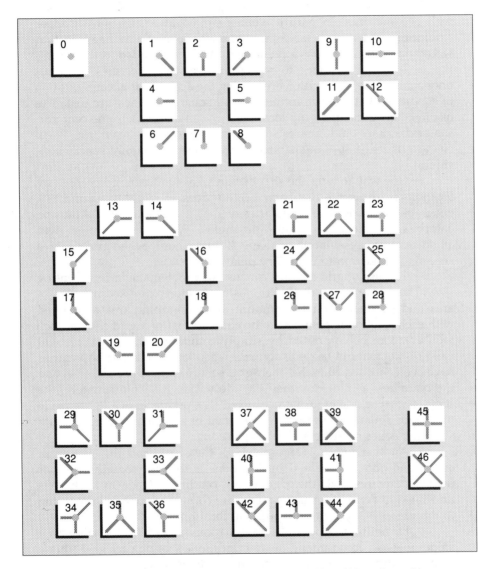

Figure 1-17 The inferred shape and size of a lineal condition within a given grid square. If lineal form is inferred as illustrated in Fig. 1-16, a lineal characteristic may assume any of 47 different configurations (gray lines) within each grid square (white).

Representing Areal Characteristics

An *areal* characteristic is one whose form encompasses two dimensions. It is a condition such as the presence of a census tract or a stand of trees that has a discernible edge and finite interior. Conditions that would otherwise be regarded as punctual or lineal may well become areal in nature when mapped at a sufficient level of resolution. At a scale of centimeters, for example, a flagpole that would normally appear as a point might well be recorded as a circle.

In using locations to represent areal characteristics, we must once again make assumptions regarding the actual spatial configuration of conditions within each location's grid square. To measure distances or directions to neighboring locations, we can adopt the same center-of-the-grid-square and nearest-neighbor conventions used for lineal characteristics. Issues of shape and size, however, require special consideration.

One way to infer the shape of an areal condition represented by locations is to assume that each location does, in fact, correspond to a geographic square. The boundaries of such a condition would thus be comprised of lateral lines and right angles. While this can often yield an adequate representation of shape, it does so only when grid squares are small in relation to boundary detail.

The use of grid squares can also be problematic in representing size. The size of an areal condition is most often expressed in terms of area and perimeter. If the amount of cartographic area associated with each location is assumed to be equal to that of a grid square, then total area can be computed by simply counting locations. This will yield results that can be quite accurate when locations along boundaries have been encoded such that their grid square corners protrude beyond the boundary as much as they intrude within it. Unfortunately, the same is not true in measuring perimeter. The perimeter of a figure made up of grid squares can greatly exceed that of a comparable figure with smoother edges regardless of grid square size.

Another way to represent both the shape and the size of an areal condition represented by locations is to infer something other than grid squares along the edges of this condition. One way to do so is illustrated in Fig. 1-18. Here, grid square corners are beveled according to the values associated with adjacent locations.

By noting the possible combinations of similar and dissimilar values that can occur among the four locations sharing each grid square corner, 15 different corner configurations can be defined. As illustrated in Fig. 1-19, each of these configurations is made up of four grid square quadrants, one from each of the adjacent locations. Note that the shape, the area, and the amount of edge associated with each of these quadrants varies from one configuration to another. To characterize the overall shape, area, or perimeter of a zone, these incremental measures would be aggregated over all the zone's locations .

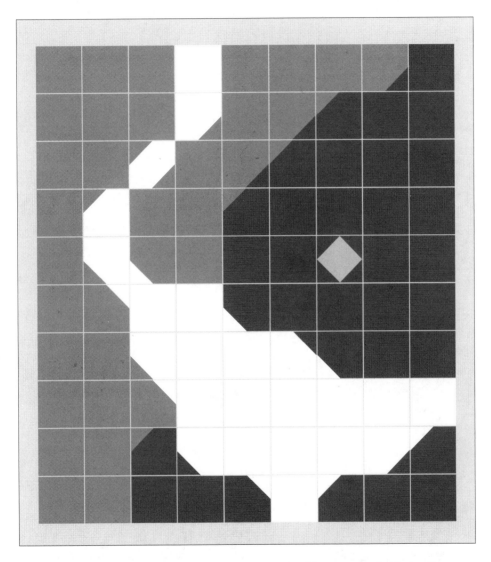

Figure 1-18 Inferring the shape and size of areal conditions represented by locations. One way to infer the spatial form of an areal condition represented by locations is to bevel grid square corners according to similarities and dissimilarities among the values (represented as shades of gray) associated with the adjacent locations.

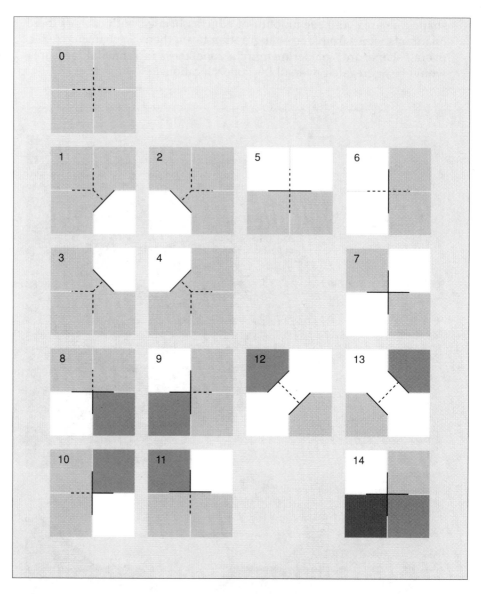

Figure 1-19   The inferred shape and size of an areal condition at the corner of a given grid square. Similarities and dissimilarities among the values associated with the four locations sharing a common grid square corner can be used to infer alternative configurations of edges between those locations. Here, each of 15 possible configurations is diagrammed by shading the adjacent quadrants of the four grid squares that share a common corner. Variations in value are indicated by variations in shading, and adjusted boundaries between locations are indicated by dashed or solid black lines. The dashed lines represent boundaries between locations of similar value, while the solid lines represent the edges at which one value meets another.

Representing Surficial Characteristics

A *surficial* characteristic is one that can be expressed such that each of its cartographic points is associated with exactly one position in a third dimension perpendicular to the cartographic plane. We refer to this third dimension as *vertical* and to those of the plane as *horizontal*. While the vertical dimension may well relate to a spatial measure such as topographic altitude or soil depth, it can also relate to nonspatial measures such as barometric pressure, travel time, or population density.

The relationship between vertical and horizontal positions that distinguishes surficial characteristics excludes forms such as spheres, cubes, and other polyhedra. Those forms will almost always associate two or more vertical positions with each horizontal position. As such, they are fully three-dimensional. While a three-dimensional form could certainly be represented as a series of layers corresponding to parallel "slices," other data structures are generally much better suited for this purpose.

The surficial distinction also excludes characteristics such as tree heights and land costs. While such characteristics do imply surfaces, these surfaces are usually discontinuous. They rise and fall abruptly at boundaries to form what amount to vertical "cliffs," thus violating the requirement that each cartographic point be associated with only one third-dimensional position. Such characteristics should generally be regarded as areal rather than surficial in nature.

Surficial characteristics can be represented with locations by using ratio or interval values to indicate position in the third dimension. If we infer from this that the entire grid square associated with each location is at a constant third-dimensional position, however, the result will be a surface of prisms like that shown in Fig. 1-20. Much like an areal form comprised of grid squares, this prismatic surface is not likely to be as useful as one in which the corners of those prisms have been beveled.

One way to achieve this effect is to set the center of each location's grid square to the elevation represented by its value, and then to interpolate the intervening elevations as shown in Fig. 1-21. Note here that the surface directly above each location's grid square is made up of eight triangular facets. Each of these can be measured in terms of characteristics such as slope, aspect, surface area, or subsurface volume. Once this has been done, the overall shape or size of a surficial condition can be characterized much as it would be for a lineal or areal condition by aggregating these incremental measurements over constituent locations.

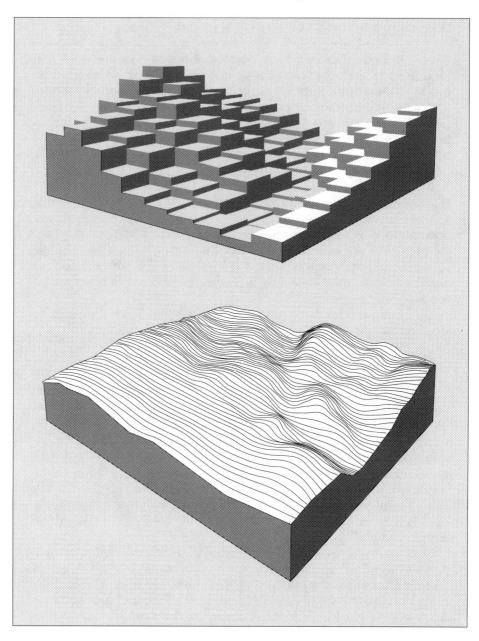

Figure 1-20 Locations in a third dimension. The form of a surficial characteristic represented by locations can be envisioned as a continuous surface on which each location is associated with a vertical position defined by its value. Such forms can be roughly approximated (above) by equating locations with grid squares in this vertical dimension. Consider, for example, the surficial form (below) of the Brown's Pond *Altitude* layer .

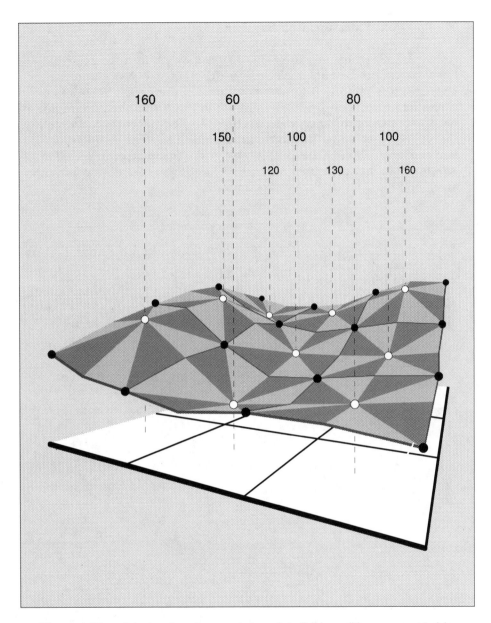

Figure 1-21 Inferring the shape and size of surficial conditions represented by locations. The spatial form of a surficial condition represented by locations can be inferred by associating the center of each location's grid square (white dot) with a third-dimensional position defined by its value (above). Each grid square corner (black dot) is at a third-dimensional position equal to the average of the values of its four diagonally adjacent neighbors, while the midpoint of each side of each grid square is at a position equal to the average of its two laterally adjacent neighbors.

The Precision of Measurements Based on Locations

In applying measurements of position or form to punctual, lineal, areal, or surficial characteristics that are represented by locations, it is important to note how this representation by locations can affect precision. As illustrated in Fig. 1-22, for example, measurements of distance and direction between punctual conditions that are represented by locations are very much affected by the resolution and orientation of the cartographic grid. This type of imprecision affects positional measures applied to locations representing lineal, areal, and surficial characteristics as well.

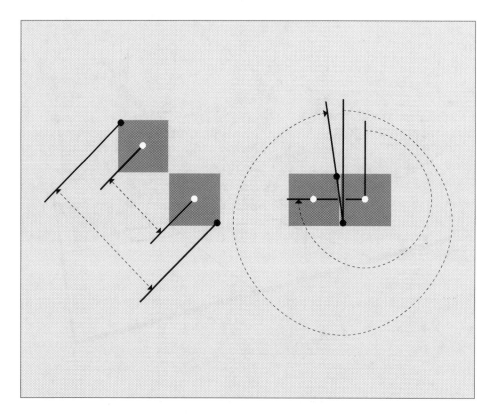

Figure 1-22 Precision of distance and directional measurements between punctual conditions represented by locations. At worst, the actual cartographic distance between the locations of two punctual conditions (black dots) may differ from the distance between the centers of the grid squares of the locations representing those conditions (white dots) by almost as much as the diagonal width of one grid square. Similarly, the angular relationship between two such conditions could conceivably differ from that of their grid square centers by almost as much as 90 degrees.

In measuring the spatial form of lineal, areal, or surficial conditions that are represented by locations, however, a second type of imprecision can have even more significant effects. This is the imprecision that results when measures of shape and size are calculated by aggregating discrete increments associated with individual locations. The implications of measuring lineal length in this manner, for example, are illustrated in Fig. 1-23. Location-based measurements of characteristics such as areal perimeter are subject to similar effects.

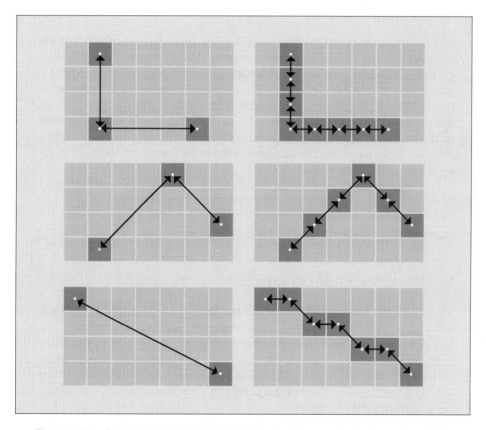

Figure 1-23 Precision of distance and directional measurements for lineal characteristics represented by locations. When the spatial form of a lineal condition (arrows) is inferred from a set of locations (darker gray squares) as illustrated in Fig. 1-16, measurements of length and orientation are affected not only by the kind of imprecision that is illustrated in Fig. 1-22 but also by incremental calculation. While lateral (top) and diagonal (center) lines can be represented in increments (right) without loss of precision, others (bottom) may be overestimated in length by as much as 8.24 percent and misrepresented in orientation by as much as 45 degrees. Significantly, this type of imprecision is independent of grid square size.

In measuring characteristics of shape and size that involve two or more dimensions, however, incremental calculations can often be quite precise. In the case of characteristics such as area, this is due in part to the compensating effects of over-estimates and under-estimates at individual locations. In the case of surficial characteristics, it is also due to the fact that third-dimensional positions can be defined with great precision (that of zonal values) and the fact that the surface implied between such positions is continuous in nature.

The Accuracy of Locational Data

In addition to the precision with which cartographic data are represented, the user of these data must also be concerned with their accuracy. Whereas precision is the degree of refinement with which a measurement can be expressed, accuracy is the degree to which such a measurement conforms to that which it represents. One way to express the accuracy of cartographic data is to describe the level of certainty with which a given value is associated with a known location. Another is to describe the certainty with which a given location is associated with a known value.

Methods of detecting, characterizing, and mitigating inaccuracies in cartographic data comprise a substantial field of study in itself. These considerations, however, need not affect the general conventions of a cartographic modeling language.

Locations on More Than One Layer

Given this understanding of the nature and significance of locations on a single layer, it remains to describe relationships between the locations on one layer and those of the other layers within the same cartographic model. Contrary to the structure implied by the diagram in Fig. 1-1, the relationship between layers and their locations is not strictly hierarchical. Each of the layers in a common model encompasses a similar set of locations.

To account for such a relationship, it is useful to extend the definition of a location from that which relates to a single layer to that which relates to an entire cartographic model. We initially defined locations as the elemental units of cartographic space that are associated with zones. To incorporate the broader view, we need only refine this definition by describing a location as illustrated in Fig. 1-24. It is the elemental unit of cartographic space that is associated with a one zone *on each of the layers* in a cartographic model. By adopting this broader definition of a location, we also extend the definitions of neighborhoods, columns, rows, and even zones to encompass more than one layer. We can now speak of "land values in the neighborhood of each location along a highway," for example, without necessarily implying that land values and highways appear on the same layer. Similarly, we might refer to "habitat diversity values (from one layer) within zones of vegetation (from another)," "visual quality ratings by ownership zone," and so on.

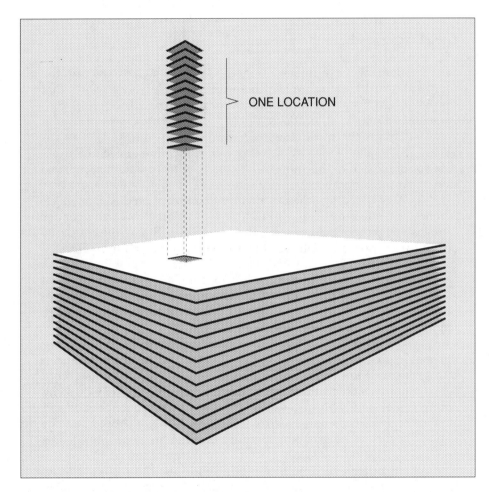

Figure 1-24 Locations on two or more layers. A location can be defined as a unit of cartographic area that represents the same geographic location on each of the layers in a cartographic model.

1-10 COORDINATES

The only major components of a location are two *coordinates*. Each of these coordinates is an integer that indicates the location's position in one of two dimensions. The first is a *column* coordinate indicating the ordinal position of the location with respect to the left edge of the cartographic grid. The second is a *row* coordinate indicating its ordinal position with respect to the lower edge of that grid. The resulting coordinate system is as shown in Fig. 1-25. On the Brown's Pond layers, both row and column coordinates range from one to 180.

Encoding Coordinates

There are a number of methods by which to record the coordinates of locations. While these vary considerably in ways that can have significant practical implications, they need not affect fundamental definition of locations.

Perhaps the most straightforward of these methods is to explicitly record the two coordinates as well as a value of each location. A major drawback to this, however, is the volume of data that must be stored. At three numbers per location, storage is highly inefficient.

Other encoding schemes attempt to improve on data storage efficiency by shifting from individual locations toward broader cartographic patterns. In general, they do so by utilizing the regularity of the cartographic grid to establish implicit rather than explicit associations between locations and their coordinates. One of the simplest and most common ways of doing this is to store the values of individual locations in an order that corresponds to an ordering of their respective positions in the cartographic grid. This reduces data storage requirements to one number (a value) per location.

To improve data storage efficiency further, other encoding schemes aggregate locations of common value and related coordinates into groups. *Run-length* encoding techniques, for example, record lateral sequences of locations that share common values. They do so by noting values and coordinates only when one such sequence ends and another begins. *Block* and *quadtree* techniques apply a similar strategy in two dimensions by grouping locations of common value into squares encoded by location and size.

Still other encoding schemes provide for the aggregation of locations that share a common value into spatial groupings that need not be rectilinear. They do so by recording sequences of locations along boundaries, isolines, networks, or other linear configurations. Among the simplest of these is *chain* encoding. Here, each sequence of locations is represented by a series of directional codes describing the relationship of each location to the next adjacent location in that sequence. More sophisticated encoding methods of this type record the coordinates of only those locations at which straight line segments change direction. These segments are then assembled into more complex graphic forms.

Raster Encoding versus Vector Encoding

These encoding methods that involve line segments are generally referred to as *vector* schemes in reference to the fact that each line segment has magnitude and direction. This is in contrast to the earlier methods, which are generally termed *raster* encoding schemes in reference to similarities between the cartographic grid and the pattern of dots in a video image.

The major differences between raster and vector encoding schemes can be expressed in terms of the two types of imprecision that are respectively illustrated in Figs. 1-22 and 1-23. The first of these relates to the resolution of the cartographic grid, while the second relates to the use of locations as discrete increments of space.

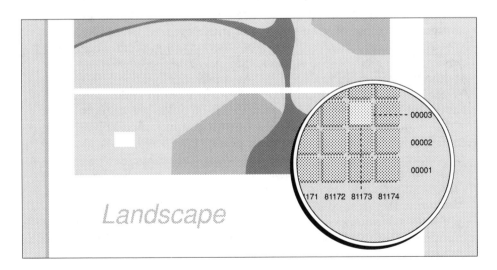

Figure 1-25   Coordinates. The coordinates of each location are two numbers that, respectively, identify its column and row positions. They are defined as consecutive integers beginning with column one and row one at the lower left corner of the cartographic grid.

As grid squares become smaller relative to the size of the conditions they are used to represent, measurements based on the centers of those squares become more and more precise. To reduce the size of the grid squares used to cover a given study area in a raster-based system requires a substantial increase in the number of squares for which locational data must be recorded. To do so in a vector-based system requires only an increase in the precision with which column and row coordinates are recorded for those locations at which boundary lines change direction. As a result, raster encoding schemes are typically used to record layers containing tens or hundreds of thousands of locations, while the number of locations on a vector-encoded layer may reach millions or billions.

Though these differences in the levels of cartographic resolution that can be achieved with raster and vector data structures are certainly significant in practical terms, they do not affect the conceptual definition of a location. Regardless of the precision with which its coordinates are defined, each location is still associated with a discrete grid square of finite size.

In contrast to this, the second type of distinction between raster and vector encoding schemes is one that can affect the fundamental definition of locations. Raster encoding clearly embodies a view in which cartographic space is quantized or broken into discrete units. This is illustrated in the first of three diagrams presented in Fig. 1-26. The same is true when vectors are used to designate locations as shown in second of those diagrams. It is not true, however, when vectors are used as shown in the third diagram. Here, though vectors begin and end at the centers of grid squares, they partition the intervening space without conforming to that grid. As a result, they appear to achieve greater precision than could have been achieved with grid squares even at the fine resolution implied by the precision of the vector coordinates.

If we adopt this third view of cartographic space, however, we can no longer rely on locations alone to represent all kinds of characteristics. While punctual conditions can still be represented as individual locations, the lineal condition that would otherwise be represented as a sequence of locations is now represented by vectors. Similarly, the areal condition that would otherwise be represented as a cluster of locations is now represented as a polygon. And the surficial condition that would otherwise be represented by locational elevations is now represented as a polyhedron.

In moving from discrete locations to vectors, polygons, and polyhedra, we move from a highly generalized unit of cartographic space to a set of units that are much more specialized. The relative advantages and disadvantages of these two cartographic data encoding

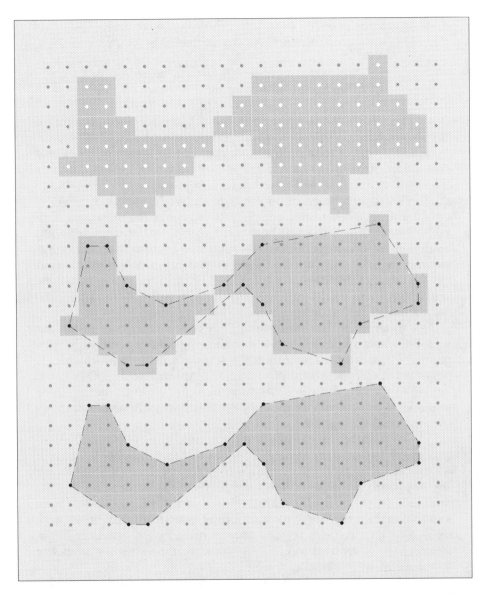

Figure 1-26 Alternative techniques for recording coordinates. One way (top) to record the coordinates of the locations within an areal zone (gray squares) is to store the zonal values of those locations in a two-dimensional array (gray and white dots) corresponding to the cartographic grid. Another (center) is to encode groups of locations within boundaries that are made up of line segments (dashed lines) defined by the coordinates of endpoints (black dots). A third technique (bottom) is to partition the cartographic plane not by way of locations at all but by way of these line segments themselves. These distinctions can be applied to techniques for the recording of punctual, lineal, and surficial as well as areal characteristics.

strategies can be assessed from both practical and conceptual points of view. One such assessment is succinctly expressed as follows:

> *Yes raster is faster, but raster is vaster,*
> *and vector just seems more correcter.*

Other assessments can be expressed somewhat less succinctly in terms of the preparation, presentation, and interpretation of cartographic data.

In terms of data preparation, the relative merits of raster and vector data structures relate to the means by which geographic facts are acquired, encoded, and stored. The data provided by earth orbiting satellites are generally in raster format, while much of what is available from both public and private mapping organizations is in vector form. Digital drafting boards (or *graphic tablet* digitizers) are also oriented toward vector data, while video input devices (or *image scanning* digitizers) are oriented more toward raster data formats. And though cartographic data at a given level of precision can generally be stored much more efficiently in vector form, this is not true when spatial variation at that level is high.

In terms of data presentation, the relative merits of raster and vector formats depend on intent. Raster graphics tend to be like photographic images and can often achieve greater realism than is possible with vector renditions. Vector graphics, on the other hand, tend to be like line drawings. As such, these are generally better suited to the kind of symbolic representation that has traditionally been associated with cartography.

And in terms of data interpretation, the relative merits of raster and vector data structures relate to the fundamental way in which each expresses relationships between *what* and *where*. Raster structures are position-oriented, while vector structures are theme-oriented. One records characteristics that are associated with locations, while the other records locations that are associated with characteristics. For this reason, raster structures are generally better suited to the interpretation of *where*, while vector structures are better suited to the interpretation of *what*.

The definition of cartographic space as a set of discrete locations can be supported by data encoding schemes that record locational coordinates in either raster or vector formats. In the case of raster data encoding, the locational construct is evident. In the case of vector encoding, however, this construct may not be obvious. To equate a line-drawn map with a grid of discrete locations, each line must be interpreted as a set of consecutive grid squares. These grid squares, however, will typically be of extremely fine resolution.

1-11 QUESTIONS

This chapter has introduced the fundamental cartographic modeling conventions that pertain to data. To review these conventions and to explore their implications, consider the following questions.

1-1 Schemes for organizing geographic data can generally be characterized according to the way in which they record theme, time, and position (or *what, when, and where*). Typically, one (such as position) is measured as a function of another (such as theme), while the third (time) is held constant. How does the structure of a cartographic model relate to this typology? How about that of an airline schedule? What about an historic account of military engagements? How could the airline schedule or the military account be expressed as cartographic models?

1-2 Imagine a map layer on which each location representing a telephone pole is set to a value of one, while all others are set to zero. Do these values relate to a nominal, ordinal, interval, or ratio scale of measurement? How about those of the Brown's Pond *Altitude* layer (and what does your answer imply about sea level)?

1-3 Why do you suppose we have used integers rather than real numbers to represent values? Could this be changed?

1-4 Surficial forms are defined in what is sometimes referred to as "two-and-a-half" dimensions. Why?

1-5 Some describe boundaries as the "natural" way of recording areal data. Is this so?

1-6 A line segment extends from a location with column and row coordinates of (0.00001, 0.00001) to another location at (0.00003, 0.00002). At this level of precision (five decimal places), is a point at position (0.000024, 0.000016) to the right or the left of that directed line segment?

1-7 Where lies the wisdom (and there is some) in a rule of thumb that advises, "Encode it in vector if it ends in an *S* and raster if it doesn't"?

Chapter 2

DATA PROCESSING

Just as methods of storing geographic data vary, so do the methods by which geographic data can be processed. These data-processing methods are also like storage methods, however, in that most can be expressed in terms of a uniform set of conventions. The data-processing conventions associated with cartographic modeling are much like those associated with manual techniques. Here, however, these conventions must be explicitly specified.

For our purposes, all data processing can be expressed in terms of individual units of processing activity and the means by which such units are combined. These are the roles respectively played by
- *operations* and
- *procedures*.

2-1 OPERATIONS

Operations are distinct and well-defined data-processing activities. The fundamental capabilities of most geographic information systems can be expressed in terms of four major types of operation, respectively associated with
- programming,
- data preparation,
- data presentation, and
- data interpretation.

Differences between these types of operations can be visualized in terms of the paths of communication among a data storage device, the processing unit of a computer, and various types of data input or output equipment. These differences are illustrated in Fig. 2-1.

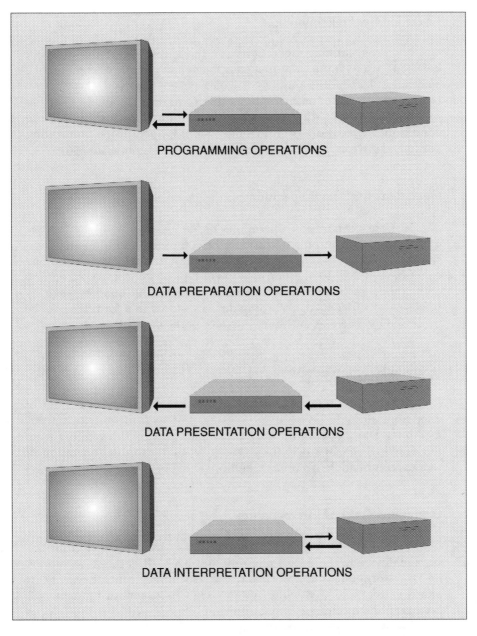

PROGRAMMING OPERATIONS

DATA PREPARATION OPERATIONS

DATA PRESENTATION OPERATIONS

DATA INTERPRETATION OPERATIONS

Figure 2-1 Communications associated with different types of operation. Paths of communication (arrows) among the input and output devices (left), the processing unit (center), and the storage facilities (right) of a geographic information system differ among operations respectively associated with programming, data preparation, data presentation, and data interpretation as indicated.

Operations for Programming

Programming operations are those that affect the dialogue between a user's input or output equipment and the processing unit of a computer. Depending on the particular system involved, these operations might provide for any of a number of routine capabilities such as initiating or terminating contact with the computer, calling for the conditional execution of previously stored operations, controlling peripheral devices, handling errors, annotating programs, and so on.

Operations for Data Preparation

Data preparation operations are associated with the flow of data from input equipment, through the computer, and ultimately into storage. Like programming operations, these too will vary from one computing system to another. They will typically encompass a variety of methods by which data can be accepted from sources such as published documents or recording devices, stored in a form that can be processed by machine, and configured for particular use.

Operations for Data Presentation

Data presentation operations are associated with a flow of data from storage device to processing unit to some sort of output medium. These operations may encompass capabilities ranging from the drawing of maps and the drafting of charts to the generation of reports, compilation of statistics, audio/video productions, and so on.

Operations for Data Interpretation

Operations that interpret data do so by cycling it from the storage device to the processing unit and back. It is these operations that transform data into information. As such, they comprise the heart of any cartographic modeling system. Our discussion of fundamental cartographic modeling capabilities in Part II will, in fact, focus exclusively on data-interpreting operations.

Data interpretation is a process in which expressions of a general nature and *potential* utility are translated into expressions of a more specialized nature and *actual* utility in a particular setting. This is a process that may involve subjective judgment as well as objective measurement. In a cartographic modeling context, it will involve operations to reassign values, superimpose layers, measure distances and directions, calculate sizes, characterize shapes, determine views, route optimal paths, simulate movement, and so on.

A more detailed view of the flow of information associated with data interpretation operations is presented in Fig. 2-2. Note here that all data interpretation is done on a layer-by-layer basis. Each operation accepts one or more *existing* layers as input and generates a *new* layer as output.

The process can be visualized as one that involves three steps. First, any existing layer that has been specified is copied from a computer's storage device into its processing unit. Next, a new layer is created from the copy(ies) according to a specified function. Ultimately, this new layer is transferred back to the storage device. When existing layers are first copied from the storage device to the processing unit, the original versions of these layers on the storage device remain intact. When the new layer is transferred back into storage, this too is done without affecting any existing layers, unless the title selected for the new layer matches that of a layer on file. In this case, the existing layer is deleted as the new layer is stored.

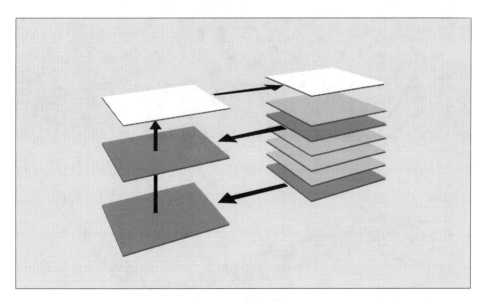

Figure 2-2   The structure of a data interpretation operation. Each data interpretation operation can be viewed as a process in which one or more existing layers (darker gray) are copied from a storage device (right) to a processing unit (left) where a new layer (white) is generated and ultimately transferred back to the storage device.

Each of the data interpretation operations is defined such that all data processing is done on a layer-by-layer basis. Whatever measurements or characterizations are applied to one location on a given layer are also applied to all others. Because of this, we can effectively describe the processing function of each operation in terms of its effect on a single, typical location. We will rely heavily on this location-oriented perspective in explaining operations and, in fact, will promote this perspective on cartographic modeling in general.

In contrast to the conventional *bird's eye* view of cartographic patterns, this *worm's eye* view from the perspective of a typical location is quite distinct. To illustrate, consider the proximity-measuring operation demonstrated in Fig. I-2. From a whole layer-oriented perspective, the output of this operation might well be described as a layer on which "Brown's Pond is surrounded by zones of uniform proximity." From an individual location-oriented perspective, on the other hand, the same new layer might instead be described as one on which "each location is characterized by its proximity to Brown's Pond." This point of view is illustrated in Fig. 2-3.

Figure 2-3 Data interpretation from the perspective of a typical location. It is often helpful to consider the effect of a data-interpreting operation not in terms of layer-wide patterns but in terms of the way in which a new value is computed for an individual location.

While the bird's-eye view does convey a more familiar graphic image, the worm's-eye view presents a description that is often more precise. It is also a form of description that will prove to be easier to build upon when operations are combined.

From the location-oriented perspective, we can distinguish three major types of data interpretation operation. As shown in Fig. 2-4, differences among the three relate to the general way in which a new value is computed for a typical location. The first type includes operations that compute a new value for each location as a function of one or more existing values associated only with that location. The second includes operations that compute each location's new value as a function of existing values within its neighborhood. And the third type of data-interpreting operation includes those that compute each location's new value as a function of existing values within a common zone.

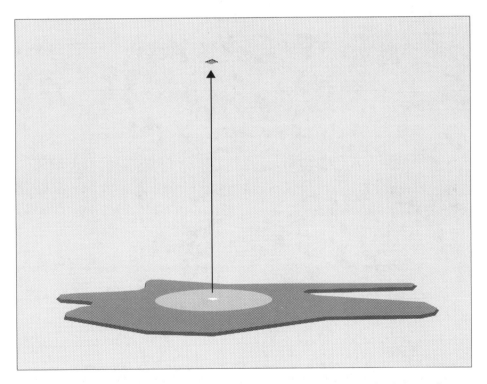

Figure 2-4 Major types of data interpretation operation. Operations that interpret data and generate new map layers as a result can be classified according to whether the new value (above) computed for any given location is calculated as a function of the location itself (white), its neighborhood (*gray*), or its zone (darker gray).

2-2 PROCEDURES

By controlling the order in which operations are performed and by using the output from one as input to another, they can be combined to form procedures. It is this ability to flexibly combine operations that makes it possible to accomplish a large set of tasks with no more than a small set of tools.

A *procedure* is any finite sequence of two or more operations that are applied to meaningful data with deliberate intent. As units of data-processing activity, procedures differ from operations in that each involves a series of steps and is highly adaptable in form.

The processing associated with a typical procedure is illustrated in Fig. 2-5. Note that procedures, like operations, accept input in the form of existing layers and generate new layers as output. Unlike operations, however, they may also employ *intermediate* layers to store results for later processing.

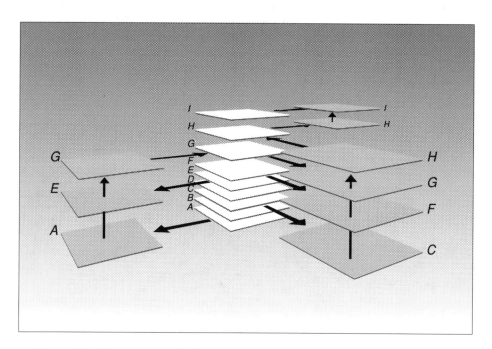

Figure 2-5 The structure of a data interpretation procedure. Procedures are typically constructed by repeatedly using a storage device (center) to retain the results of one operation for later processing by another. Here, a sequence of three operations is used to
- generate an intermediate layer *G* from existing layers *E* and *A*,
- generate another intermediate layer *H* from *G*, *F*, and *C*, and
- finally generate a new layer *I* from *H*.

Without this ability to store intermediate layers, the logical structure of a procedure would be limited to the linear sequence in which operations are specified. With it, more complex structures are possible. Most procedures embody a logical structure in which two or more independent sequences of operations eventually converge. This structure can be represented as a hierarchy of layers in which each layer created is related to the layer(s) from which it was generated by way of an operation. A simple example of this structure is illustrated in Fig. 2-6. Note that it can be equated with the structure of an algebraic expression.

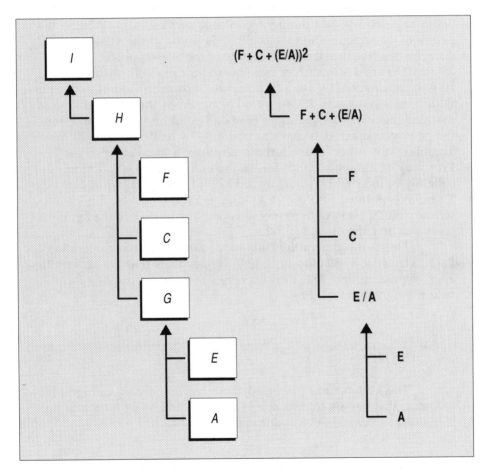

Figure 2-6 Diagrammatic representation of a data interpretation procedure. The logical structure (left) of the procedure illustrated in Fig. 2-5 is shown to be similar to the structure (right) of a typical algebraic expression.

The ability to combine data-interpreting operations relies not only on the compatibility of input and output formats, but also on the fundamental way in which map layers convey cartographic data. As indicated earlier, this is done with numerical values that are explicitly associated with individual locations.

In traditional cartography, the content of a map is typically conveyed by way of lines, symbols, stylized icons, and similar graphic devices. To record the geographic characteristics of individual locations, these devices often rely on an ability to view each location in its spatial context. This is especially true in recording geometric characteristics such as distance, direction, size, shape, and movement. Consider, for example, the conventional cartographic image in Fig. 1-3. Select any point within the light gray area on this map. What is the nature of its surrounding topography? Is it east of the village center? Can you describe the size and the shape of the nearest pond?

The answer to each of these questions can easily be found (or at least approximated) by eye. By computer, however, we must contend with what amounts to a form of myopia. From the perspective of a machine that can "see" no more than one point on the map at a time, the point you selected is characterized only by its designation as gray. To infer any other characteristic requires a broader perspective. Proximity to Brown's Pond, for example, is a characteristic that can certainly be inferred by visual inspection of the map shown in Fig. 1-3. To be incorporated into a cartographic model, however, this characteristic would first have to be measured and then stored in the form of a layer like that shown in Fig. I-2.

This is the general function of all data-interpreting operations. Each detects facts, relationships, and/or meanings that are implicit in a set of existing data and then expresses them in the explicit form of values assigned to locations.

2-3 QUESTIONS

This chapter has presented those conventions of cartographic modeling that pertain to data processing. The following questions further examine this data-processing construct.

2-1 The data-processing conventions presented in this chapter are those of a system designed not for *querying* but for *modeling*. What is the distinction between these activities, and how is it reflected in this data-processing construct?

2-2 How would the *ThisMuchMore* layer shown in Fig. I-4 be described from contrasting worm's and bird's eye perspectives?

2-3 It is often useful to be able to modify the normal linear or hierarchical structure of a procedure by skipping or repeating specified sequences of operations conditionally. If this capability were to be expressed as an operation, to which of the four major types of operation would it belong? How about an operation to change a layer's title, its resolution, its orientation, or its labels?

2-4 Chapter 1 made passing reference to the accuracy of cartographic data. Suppose that each of the locations on a given map layer could be characterized in terms of the likelihood that its value is accurate. Given the data-processing structure outlined in this chapter, how could that likelihood-of-accuracy (or probability-of-error) figure be used to estimate error propagation when the layer is used in a procedure?

Chapter 3

DATA-PROCESSING
CONTROL

Having adopted sets of conventions relating to data and data processing, we must also establish a set of conventions for data-processing control. This is a matter of specifying operations, indicating the data to which they are to be applied, and designating the order in which they are to be performed.

The way in which this control is actually exercised may well vary from one geographic information system to another. Nonetheless, a common set of conventions for data-processing control can be expressed by adopting a uniform notational format. Here, we employ a form of notation or language that resembles both algebra and English. It is longhand notation designed for ease of instruction, translation, and modification.

The two major elements in this notation are respectively associated with operations and procedures, These include

- *statements* and
- *programs.*

3-1 STATEMENTS

A *statement* is the notational representation of an operation. The function of a statement is much like that of what are termed *instructions*, *steps*, and *commands* as well as *statements* in other computing contexts. The form of a statement, however, is more like a declarative English sentence.

Each statement is an ordered sequence of letters, numerals, symbols, and/or blank spaces. Throughout this text, all statements or portions thereof are presented in *small, italic* type. Statements are specified such that consecutive characters read from left to right and continue from each line of text to indented lines below as necessary. The following, for example, is a typical cartographic modeling statement.

AverageCost   = *LocalMean of YourCost and MyCost*
                     and HerCost and HisCost

This statement calls for the generation of a new layer entitled *AverageCost* by averaging the values of existing layers *YourCost* , *MyCost*, *HerCost*, and *HisCost* on a location-by-location basis.

Strings of consecutive letters, numerals, and/or symbols within a statement are separated from one another by one or more blank spaces. These strings of characters form
- *subjects,*
- *modifiers,* and
- *objects.*

Every statement begins with its *subject*, the title of a new map layer. This is the layer to be created when the statement is executed. *AverageCost* is the subject in the statement cited above.

Certain statements may also include one or more *modifiers*. These correspond in both form and function to conventional prepositions, adverbs, adjectives, nouns, or marks of punctuation. Modifiers in the statement cited above include *=*, *of,* and *and*.

The modifiers of a statement may also refer to *objects* representing data or processing options. Objects in the statement cited earlier, for example, include *LocalMean*, *YourCost* , *MyCost*, *HerCost*, and *HisCost* .

Objects may be layer titles, nouns, adverbs, or numerals. They may also be special numerical codes including
- *-0* ,
- *++,*
- *--* , and
- *...* .

-0 is used to represent the null value. *++* and *--* are used to represent,

respectively, the highest and lowest numbers available on a given computing system. And ... refers to all numbers between the one immediately preceding (or -- if there is none) and the one immediately following (or ++ if there is none). *1 ... 5*, for example, would refer to what might otherwise be specified as *1 2 3 4 5*, while *1 ...* refers to all numbers that are greater than or equal to one, and *... 1* refers to all numbers that are less than or equal to one. Our use of the term *numeral* throughout the text is meant to include these special codes.

It is often necessary to represent portions of a statement as generic variables without making any specific reference to a particular subject, modifier, or object. This is done by specifying those portions of the statement as strings of all-uppercase characters. These strings are to be replaced by actual titles, numerals, and so on when a statement is specified for execution. Each string is intended to suggest the nature of that which it represents. For example, a generalized expression of the *LocalMean* statement cited earlier might be given as

NEWLAYER = *LocalMean of FIRSTLAYER and SECONDLAYER*
and THIRDLAYER and FOURTHLAYER

It is often helpful to be able to generalize statements in other ways as well. To designate portions of a statement that are optional, for example, each is enclosed in *[*rectangular brackets*]*. The *LocalMean* statement, for example, might be presented as

NEWLAYER = *LocalMean of FIRSTLAYER and SECONDLAYER*
[and THIRDLAYER] [and FOURTHLAYER]

to indicate that only two existing layers are mandatory.

Still further generalization is possible as well through use of the term *etc.* This indicates that any immediately preceding notation in rectangular brackets may be repeated. For example,

NEWLAYER = *LocalMean of FIRSTLAYER and SECONDLAYER*
[and NEXTLAYER] etc.

would indicate that the operation can be applied to any number of existing map layers in addition to the initial two. Repetition of a phrase not followed by *etc.* would effectively cancel the earlier version.

Statements and operations may be referred to in even more generalized form by eliminating certain terms and phrases altogether. Note, for example, how we have already referred to "the *LocalMean* statement" above.

The fundamental capabilities of cartographic modeling can be expressed in terms of variations on four major operations given as

- *FocalFUNCTION*,
- *IncrementalFUNCTION*,
- *LocalFUNCTION*, and
- *ZonalFUNCTION*.

The statements for these operations are specified as follows:

```
NEWLAYER     = FocalFUNCTION of FIRSTLAYER
                 [at DISTANCE] etc.   [by DIRECTION] etc.
                 [spreading [in FRICTIONLAYER]
                         [on SURFACELAYER]
                           [through NETWORKLAYER] ]
                 [radiating [on SURFACELAYER]
                         [from TRANSMISSIONLAYER]
                         [through OBSTRUCTIONLAYER]
                         [to RECEPTIONLAYER] ]


NEWLAYER     = IncrementalFUNCTION [of FIRSTLAYER] [on SURFACELAYER]


NEWLAYER     = LocalFUNCTION of FIRSTLAYER [and NEXTLAYER] etc.


NEWLAYER     = ZonalFUNCTION of FIRSTLAYER [within SECONDLAYER]
```

In each case,

- *FUNCTION* is a name such as *Sum*, *Volume*, or *Proximity* that specifies a processing option,
- *DISTANCE* and *DIRECTION* are numerals setting parameters of this processing option,
- *FIRSTLAYER*, *SECONDLAYER*, *NEXTLAYER*, *FRICTIONLAYER*, *SURFACELAYER*, *NETWORKLAYER*, *TRANSMISSIONLAYER*, *OBSTRUCTIONLAYER*, and *RECEPTIONLAYER* are titles identifying the existing layer(s) to which this processing is to be applied, and
- *NEWLAYER* is the title of a new map layer.

Fuller descriptions of these statements are presented in the Appendix.

3-2 PROGRAMS

A ***program*** is the notational representation of a procedure. It is a sequence of statements specified such that each begins on the first line of text below the one (if any) before it.

The order in which statements are specified in a program generally indicates the order in which corresponding operations are to be performed. Variations in this order can be specified, however, by way of programming operations that cause designated sequences of statements to be skipped, inserted, or repeated. In either case, it is through the titles of the layers that are processed by these operations that the logical structure of a procedure is defined. To illustrate this, consider the following set of consecutive statements:

TotalCost	= *LocalSum of YourCost and MyCost*	
TotalCost	= *LocalSum of TotalCost and HerCost*	
TotalCost	= *LocalSum of TotalCost and HisCost*	
AverageCost	= *LocalRatio of TotalCost and 4*	

If *LocalSum* and *LocalRatio* respectively refer (as they do) to addition and division functions, then the effect of this procedure can be equated with that of operation

AverageCost   = *LocalMean of YourCost and MyCost and HerCost and HisCost*

where *LocalMean* refers to a similar function computing averages.

To represent programs in generalized form, rectangular brackets and *etc.* may be used as they are within statements. To replicate

NEWLAYER   = *LocalMean of FIRSTLAYER and SECONDLAYER*
         [and NEXTLAYER] etc.

for example, a (tedious but illustrative) program might be specified as follows:

HowMuch	= *LocalSum of FIRSTLAYER and SECONDLAYER*	
HowMany	= *LocalSum of 1 and 1*	
[*HowMuch*	= *LocalSum of HowMuch and NEXTLAYER*	
HowMany	= *LocalSum of HowMany and 1] etc.*	
NEWLAYER	= *LocalRatio of HowMuch and HowMany*	

3-3 QUESTIONS

This chapter has outlined the data-processing control conventions associated with cartographic modeling. The following are questions that reconsider these conventions.

3-1 This chapter has described a particular medium that is largely incidental to the more general message it is used to convey. How could the (sometimes quite tedious) longhand form of a statement be abbreviated? How about that of a program?

3-2 How could this statement format be translated into a less verbal and more graphic mode of interaction?

3-3 The statement format presented calls for specification of
- a dependent variable (the title of a new layer), followed by
- the function generating that variable (the name of an operation),
- the function's independent variable(s) (existing layer title(s)), and
- any function modifiers.

What are some alternatives to this sequence, and what are their implications for the way in which an operation is visualized by its user?

3-4 How does the cartographic model of a particular study area (such as Brown's Pond) differ from the cartographic model of a geographic phenomenon (such as soil erosion) that is not site specific?

Part II

CARTOGRAPHIC MODELING CAPABILITIES

Chapter 4

CHARACTERIZING
INDIVIDUAL LOCATIONS

Having outlined the structure of a cartographic modeling language in terms of conventions, the substance of this language can now be defined in terms of its capabilities. These capabilities are embodied in the functions performed by individual operations and the ways in which operations are combined into procedures. The following three chapters introduce the fundamental capabilities of a cartographic modeling system by describing data interpretation operations that characterize
- individual locations,
- locations within neighborhoods, and
- locations within zones,

respectively. They offer what amounts to a systematic introduction to a set of tools. This toolbox will eventually be equipped with several dozen variations on four major operations. To keep them within easy reach, a summary description of each operation is also presented in the Appendix.

The first major group of data-interpreting operations includes those that compute a new value for each location on a layer as a function of existing data explicitly associated with that location. The data to be processed by these operations may include any one or more of the zonal values associated with each location. This gives rise to a useful distinction between
- those operations that generate new values as a function of existing values on a single layer, and
- those that do so as a function of existing values on two or more layers.

4-1 FUNCTIONS OF A SINGLE VALUE

The first group of operations characterizing individual locations includes those that compute a new value for each location as a specified function of that location's value on a single existing layer. The general process is illustrated in Fig. 4-1. Among the operations in this group are

- *LocalRating*,
- *LocalMaximum, LocalMinimum, LocalSum, LocalDifference, LocalProduct, LocalRatio,* and *LocalRoot*, and
- *LocalSine, LocalCosine, LocalTangent, LocalArcSine, LocalArcCosine,* and *LocalArcTangent*.

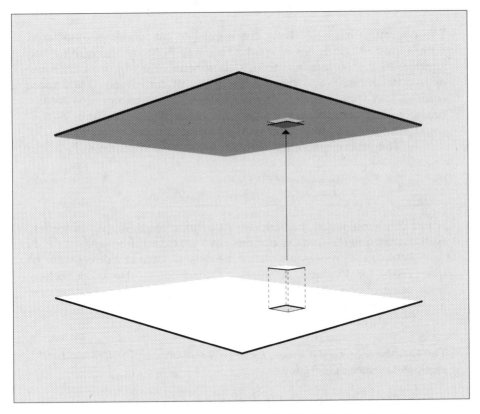

Figure 4-1 Functions of single values associated with individual locations. Operations *LocalArcCosine, LocalArcSine, LocalArcTangent, LocalCosine, Local-Difference, LocalMaximum, LocalMinimum, LocalProduct, LocalRating, LocalRatio, LocalRoot, LocalSine, LocalSum,* and *LocalTangent* can all be used to compute a new value (above) for each location as a specified function of that location's value on an existing map layer (below).

The *LocalRating* Operation

One of the most basic interpretive operations in any data-processing system is that which provides for the explicit assignment of new values to existing variables. The process is variously referred to as *recoding, reclassifying,* or transforming through *look-up tables.* In the present context, it involves generating a new map layer on which each location is set to a value that has been assigned to its zone on an existing layer. This is done by way of an operation called *LocalRating.*

To illustrate the use of this operation, consider the layer of land development presented in Fig. 1-12. Fig. 4-2 shows a new layer generated from this *Development* layer with an operation given as

Mobility   = *LocalRating of Development with 12 for 1 2 with 36 for 0 3 4 5*

The operation has effectively distinguished all locations associated with either of two types of road. This was done by aggregating the zones of *Development* into two groups: those with values of one (*MajorRoads*) or two (*MinorRoads*), and those with values of zero (*VacantLand*), three (*Houses*), four (*PublicBuildings*), or five (*Cemeteries*). The two groups are ultimately represented as new zones with values of 12 (*Road*) and 36 (*NonRoads*), respectively, on a layer entitled *Mobility.*

The generic statement for this operation is specified as

NEWLAYER   = *LocalRating of FIRSTLAYER*
                [with NEWVALUE for FIRSTZONES] etc.

For a full explanation, consult the Appendix. In doing so, however, realize that other *LocalRating* options have yet to be introduced.

LocalRating represents one of the most flexible and direct ways of interpreting local values. As such, it finds frequent use in connection with attempts to aggregate, isolate, weight, or otherwise express judgment concerning local characteristics.

The *LocalMaximum, LocalMinimum, LocalSum, LocalDifference, LocalProduct, LocalRatio,* and *LocalRoot* Operations

Often, the relationship between a set of values representing zones and a set of new values to be assigned to those zones is systematic enough to be expressed in the form of a mathematical function. Suppose, for example, that existing values of 15, 25, 35, and 45 are to be translated into new values of 30, 50, 70, and 90, respectively. This could

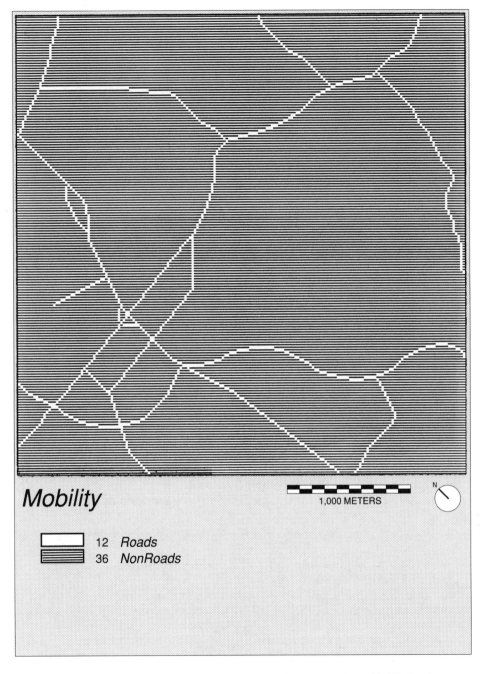

Figure 4-2 A map layer created using the *LocalRating* operation. *Mobility* is a layer depicting those locations in the Brown's Pond study area that lie along major or minor roads.

be done by specifying a *LocalRating* operation with phrases such as *with 30 for 15*, *with 50 for 25*, and so on. To do this for a large number of values, however, would quickly become impractical. A better approach might simply be to multiply each value by two.

To provide this type of capability, we introduce a set of operations that, like *LocalRating*, transform the values of an existing layer on a zone-by-zone basis. Unlike *LocalRating*, however, these operations do not require that each new value be explicitly specified. Instead, each zone's new value is computed as a mathematical function of its existing value. The use of such mathematical functions generally requires that these values be defined over ratio, interval, or at least ordinal scales of measurement.

One such group of operations includes those that transform the values of an existing layer by computing a simple algebraic function of each value and a specified constant. To illustrate this, consider the following transformation of the *Altitude* layer shown in Fig. 1-9:

Contouring	= *LocalRatio of Altitude and 10*
Contouring	= *LocalProduct of Contouring and 10*
Contouring	= *LocalDifference of Altitude and Contouring*
Contouring	= *LocalRating of Contouring*
	with 4 for 6 with 3 for 7 with 2 for 8 with 1 for 9

Here, each *Altitude* value is divided using an operation called *LocalRatio*, then multiplied by *LocalProduct*, and subtracted with *LocalDifference*. The result, shown in Fig. 4-3, is a *Contouring* layer on which each location's value indicates the final digit of its topographic elevation in meters.

Maxima, minima, sums, and roots can also be computed for values and constants by way of operations respectively given as *Local-Maximum*, *LocalMinimum*, *LocalSum*, and *LocalRoot*.

The statements for these operations are specified as

NEWLAYER = *LocalFUNCTION of FIRSTLAYER and SECONDLAYER*

where *FIRSTLAYER* is a title and *SECONDLAYER* a numeral, or *FIRSTLAYER* is a numeral and *SECONDLAYER* a title.

These operations typically find use in converting values from one set of measurement units to another. Note, however, that *LocalProduct*, *LocalRatio*, and *LocalRoot* generally produce meaningful results only when applied to values that relate to a ratio scale of measurement. Operations *LocalSum* and *LocalDifference* can be applied to interval as well as ratio values, while both *LocalMinimum* and *LocalMaximum* can be applied to values that relate to ordinal, interval, or ratio scales.

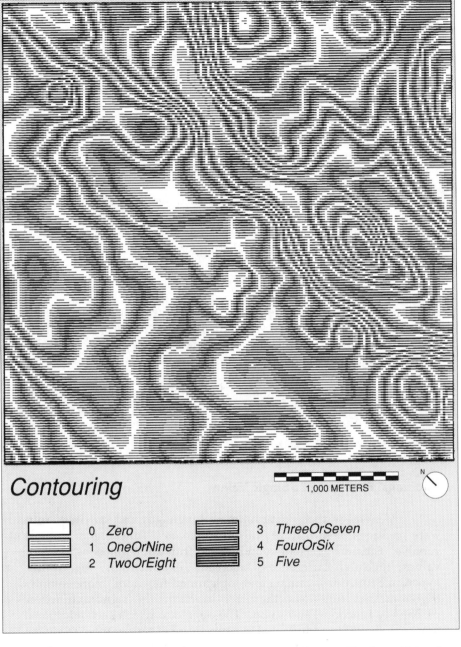

Contouring

1,000 METERS

N

0 *Zero*
1 *OneOrNine*
2 *TwoOrEight*
3 *ThreeOrSeven*
4 *FourOrSix*
5 *Five*

Figure 4-3   A map layer created using the *LocalDifference*, *LocalProduct*, and *Local-Ratio* operations. *Contouring* is a layer indicating the final digit of each location's value on the Brown's Pond *Altitude* layer.

The *LocalSine, LocalCosine, LocalTangent, LocalArcSine, LocalArcCosine,* **and** *LocalArcTangent* **Operations**

Local values can also be transformed with trigonometric functions. In Fig. 4-4, for example, is a layer on which each location has been set to a function of its pond proximity value on the *ThatFar* layer shown in Fig. I-3 as follows:

ThatIntensity	=	*LocalRatio of ThatFar and 20*
ThatIntensity	=	*LocalTangent of ThatIntensity*
ThatIntensity	=	*LocalRatio of 10000 and ThatIntensity*

Sines, cosines, arc sines, arc cosines, and arc tangents can be applied in a similar manner by way of operations *LocalSine, LocalCosine, LocalArcSine, LocalArcCosine,* and *LocalArcTangent,* respectively.

The statement for each of these operations is specified as

NEWLAYER = *LocalFUNCTION of FIRSTLAYER*

where *FIRSTLAYER* is the title of an existing layer whose values represent the angles, in degrees, to which the functions are to be applied.

Trigonometric functions are generally used in cartographic models to deal with angular measurements such as those associated with compass directions, topographic relationships, lines of sight, and cartographic projection systems. These functions are also used to characterize cyclical variations over time and/or space such as those associated with climatic patterns, geomorphological processes, and biological or social activity.

Additional Functions of a Single Value

General-purpose programming languages and statistical software packages often feature other mathematical functions of a single variable. These may include rounding and truncation functions, absolute values, modular differences, factorials, exponentials, logarithms, hyperbolic functions, gamma and log-gamma functions, more complex trigonometric functions, the error function, the signum function, probability functions, and so on. The degree to which such additional functions should be represented in a repertoire of cartographic modeling operations generally depends on how frequently those functions are likely to be used and how well they can be replicated, if at all, by combining more primitive operations.

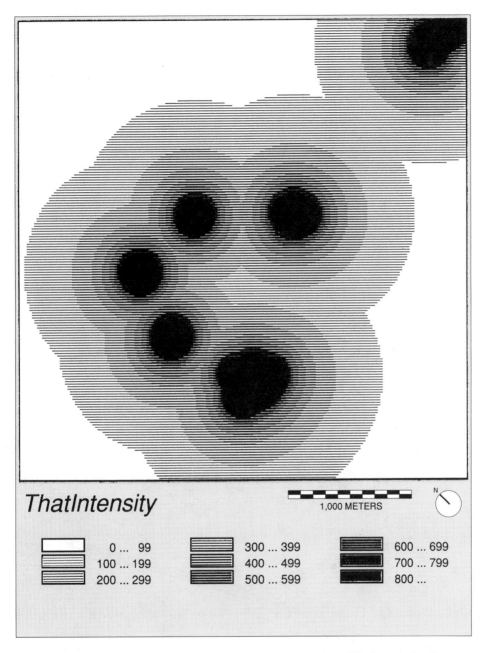

Figure 4-4 A layer created using the *LocalTangent* operation. *ThatIntensity* is a layer indicating the magnitude of some (unspecified) quantity in the Brown's Pond study area that dissipates with the cotangent of distance from selected ponds. Note that each shading pattern represents not just one zone but a range of intensity levels.

4-2 FUNCTIONS OF MULTIPLE VALUES

The second group of operations characterizing individual locations includes those that compute a new value for each location as a specified function of its existing values on two or more layers. This is done as illustrated in Fig. 4-5. The process is sometimes referred to as *compositing, overlaying,* or *superimposing* maps. Among these operations are

- *LocalRating,*
- *LocalCombination,*
- *LocalVariety,*
- *LocalMajority* and *LocalMinority,*
- *LocalMaximum* and *LocalMinimum,* and
- *LocalSum, LocalDifference, LocalProduct, Local Ratio, LocalRoot, LocalArcTangent,* and *LocalMean.*

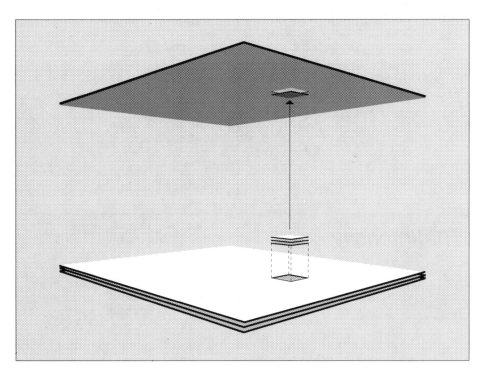

Figure 4-5 Functions of multiple values associated with individual locations. Operations *LocalArcTangent, LocalCombination, LocalDifference, LocalMajority, LocalMaximum, LocalMinimum LocalMinority, LocalMean, LocalProduct, LocalRating, LocalRatio, LocalRoot, LocalSum,* and *LocalVariety* can all be used to compute a new value for each location as a specified function of that location's values on two or more existing map layers.

The *LocalRating* Operation

To characterize locations in terms of values from two or more layers calls for processing capabilities that can be viewed as extensions of the operations so far described. Consider, for example, a version of *LocalRating* in which the new values to be assigned to zones are specified not as constants but as variables. This can be done by setting each location in a designated zone to its existing value on a specified layer.

An example is shown in Fig. 4-6. Here, an operation given as

OpenDevelopment = *LocalRating of Vegetation with Development for 0 with 6 for 1 ... 3*

has been used to generate a new layer entitled *OpenDevelopment* on which *Vegetation* zones one, two, and three (*HardWoods*, *SoftWoods*, and *Mixed-Woods*) are set to a value of six, while each location in zone zero (*Open-Land*) is set to that location's value on the existing *Development* layer.

This form of the *LocalRating* operation is specified as

NEWLAYER = *LocalRating of FIRSTLAYER*
  [with SECONDLAYER for FIRSTZONES] etc.

where *SECONDLAYER* is the title of an existing layer from which new values are to be taken on a location-by-location basis.

Another variation on *LocalRating* that can be used to characterize locations in terms of their values on two or more layers is given as

NEWLAYER = *LocalRating of FIRSTLAYER [and NEXTLAYER] etc.*
  [with NEWVALUE for FIRSTZONES [on NEXTZONES] etc.] etc.

This provides for the assignment of new values not to individual zones but to locations where zones coincide. Fig. 4-7 shows an example in which *Altitude* and *Vegetation* are used as follows:

WindExposure = *LocalRating of Altitude and Vegetation*
  with 0 for 290 ... on 0 with 1 for 290 ... on 1 ... 3
  with 2 for ... 289 on 0 1 3 with 3 for ... 289 on 2

to identify particular combinations of topographic elevation and forest cover that vary in wind exposure.

This form of *LocalRating* is especially important for those situations in which rules of combination cannot be expressed in the generalized form of a mathematical function. Such is often the case when values to be combined relate to no more than a nominal scale.

OpenDevelopment

1,000 METERS

N

	0	OpenVacantLand
	1	MajorRoads
	2	MinorRoads
	3	Houses
	4	PublicBuildings
	5	Cemeteries
	6	Woods

Figure 4-6 A map layer created using the *LocalRating* operation to assign values from one existing map layer to another. *OpenDevelopment* is a layer on which each location in the Brown's Pond study area is set to a value indicating whether or not it is wooded and, if not, what type of development it contains.

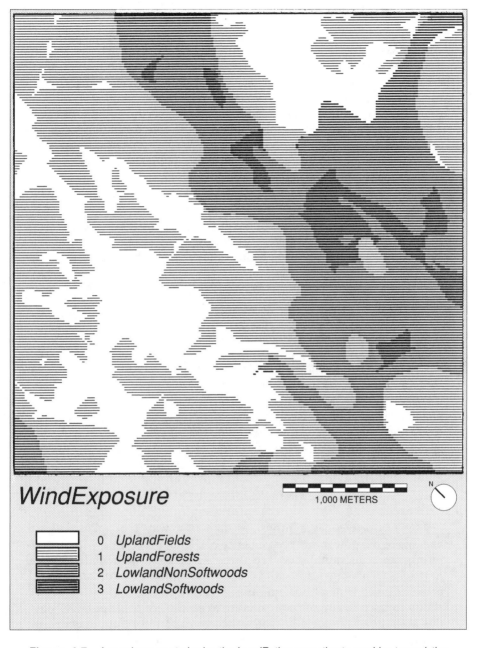

WindExposure

1,000 METERS

N

- ⬜ 0 *UplandFields*
- 🮒 1 *UplandForests*
- 🮕 2 *LowlandNonSoftwoods*
- ⬛ 3 *LowlandSoftwoods*

Figure 4-7 A map layer created using the *LocalRating* operation to combine two existing layers. *WindExposure* is a layer on which each location in the Brown's Pond study area is set to a value distinguishing uplands with and without tree cover from lowlands with and without softwood tree cover.

The *LocalCombination* **Operation**

With the *LocalRating* operation, any new value can be assigned to any local combination of zones. As indicated earlier, however, there are practical limitations on the amount of data that can be handled in this manner. To combine just two layers with five zones apiece could require as many as five times five or 25 explicit specifications. And with a third layer of five zones, this would grow to five times five times five or 125.

Seldom, however, will all possible combinations of the zones on two or more map layers actually occur in a given study area. If we combine two layers of five zones apiece, we may well find that only 10 or 15 of the 25 possible combinations are actually present. In light of this, it is often useful to generate a map layer that combines the local values of two or more existing layers by simply assigning a unique new value to each combination of existing values that actually occurs. This is the function of operation *LocalCombination*.

An example of *LocalCombination* is presented in Fig. 4-8. Here, an operation given as

LandCover   = *LocalCombination of Water and Vegetation and Development*

has been used to generate a layer on which each location is characterized in terms of its particular combination of *Water*, *Vegetation*, and *Development* zones. Note that, of the 96 (four times four times six) combinations possible, only 28 actually occur. Nowhere in the Brown's Pond study area, for example, do we find wooded ponds, houses on wetlands, or major roads through softwoods.

The statement for this operation is specified as

NEWLAYER   = *LocalCombination of FIRSTLAYER [and NEXTLAYER] etc.*

LocalCombination synthesizes data in a way that takes full advantage of the power and convenience of digital processing without suffering the constraints that are often associated with mathematical functions. This operation can be used with nominal, ordinal, interval, or ratio data; it demands no subjective judgment; and it retains an ability to trace the inputs contributing to each piece of output. As such, this operation is especially well suited to exploratory applications intended to investigate a particular study area. Unfortunately, its specificity to the combination of conditions present in that particular study area generally makes *LocalCombination* unsuitable for use in generic models to be applied elsewhere.

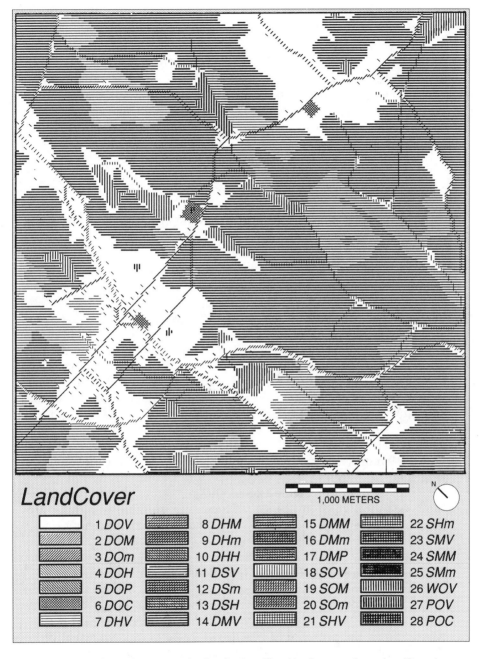

Figure 4-8 A map layer created using the *LocalCombination* operation. *LandCover* is a layer indicating the combination of *Water*, *Vegetation*, and *Development* zones at each location in the Brown's Pond study area. Each label is an acronym for three such zones.

The *LocalVariety* Operation

Another operation that can be applied to nominal as well as ordinal, interval, or ratio data is *LocalVariety*. This operation generates a new value for each location indicating how much diversity exists among the values that are associated with that location on a specified set of existing layers. If values of 17, 12, 14, 13, 13, 15, and 12 were to be associated with a given location, for example, its *LocalVariety* value would be five.

The statement for this operation is specified as

NEWLAYER = *LocalVariety of FIRSTLAYER [and NEXTLAYER] etc.*

An example of *LocalVariety* is presented in Fig. 4-9. Here, an operation given as

TypesOfScore = *LocalVariety of ThisScore and ThatScore and HisScore and HerScore*

has been used to indicate the degree of agreement or disagreement among four assessments of land use suitability in the Brown's Pond study area. In Fig. 4-10 is a layer entitled *ThisScore* on which each location is set to a value of zero, 10, 20, or 30. These values respectively indicate those areas within the Brown's Pond study area that are judged by one individual to be of very low, low, high, and very high potential for the siting of new houses. In Figs. 4-11 through 4-13 are layers entitled *ThatScore*, *HisScore*, and *HerScore* on which similar values represent the site suitability judgments of other individuals. The *ScoreDiversity* layer in Fig. 4-9 represents one way of summarizing these four expressions of judgment.

Note that this method of combining values is not affected by their quantitative significance. Here, site suitability ratings are treated not as levels of magnitude but only as nominal designations.

The *LocalMajority* and *LocalMinority* Operations

The ability of *LocalVariety* to process nominal as well as ordinal, interval, and ratio values relates to the fact that this operation is sensitive only to the presence or absence of values. It is not sensitive to their magnitude. The same is true of an operation called *LocalMajority*. This is an operation that assigns to each location a new value equal to whichever of a set of existing values occurs most often at that location.

Figure 4-9 A map layer created using the *LocalVariety* operation. *ScoreDiversity* is a layer on which each location in the Brown's Pond study area is set to a value indicating how many different land development suitability ratings have been applied to that location by four different individuals.

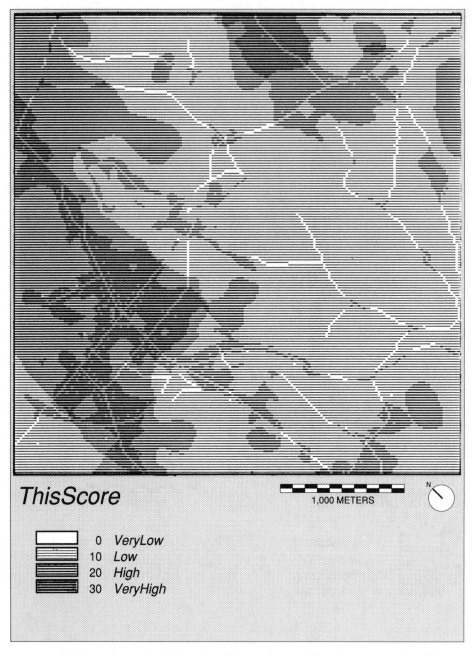

Figure 4-10 A map layer reflecting one individual's view of land development potential. *ThisScore* is a layer of the Brown's Pond cartographic model on which each location's value reflects one assessment of its suitability for residential development.

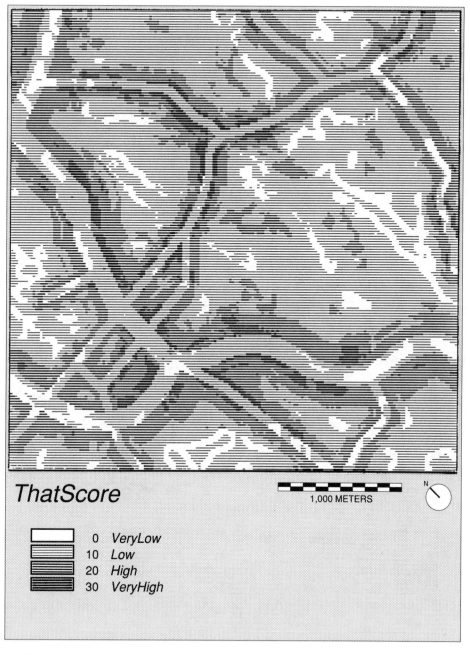

Figure 4-11 A map layer reflecting a second view of land development potential. *That-Score* is a layer of the Brown's Pond cartographic model on which each location's value reflects another assessment of its suitability residential development.

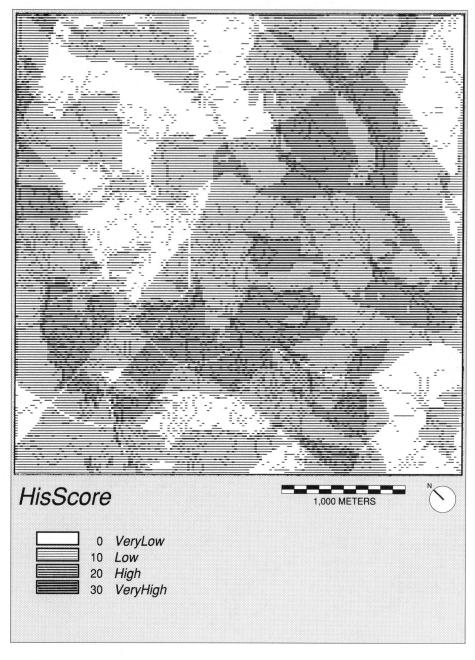

HisScore

1,000 METERS

N

	0	VeryLow
	10	Low
	20	High
	30	VeryHigh

Figure 4-12 A map layer reflecting a third view of land development potential. *HisScore* is a layer of the Brown's Pond cartographic model on which each location's value reflects another assessment of its suitability for residential development.

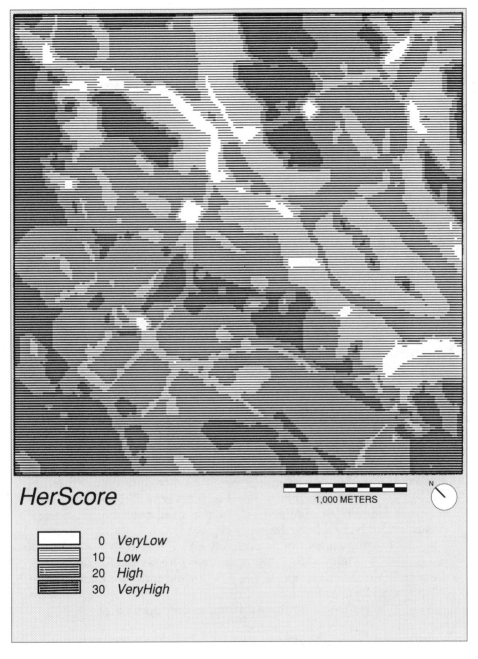

HerScore

1,000 METERS

N

	0	*VeryLow*
	10	*Low*
	20	*High*
	30	*VeryHigh*

Figure 4-13 A map layer reflecting a fourth view of land development potential. *HerScore* is a layer of the Brown's Pond cartographic model on which each location's value reflects another assessment of its suitability for residential development.

LocalMinority is a similar operation. In this case, however, it is the value occurring least frequently that is ultimately assigned. If two or more existing values share the most- or least-frequent distinction, a null value is assigned.

The statements for *LocalMajority* and *LocalMinority* are specified as

NEWLAYER   = *LocalFUNCTION of FIRSTLAYER [and NEXTLAYER] etc.*

In Fig. 4-14 is an example of their use. Here, a new layer entitled *PredominantScore* has been generated as follows:

PredominantScore   =   *LocalMajority of ThisScore and ThatScore*
                           and HisScore and HerScore

PredominantScore summarizes the four values associated with each location on *ThisScore*, *ThatScore*, *HisScore*, and *HerScore* by selecting whichever score, if any, occurs most often.

Relative to operations *LocalRating*, *LocalCombination*, and *LocalVariety*, this method of summarizing values represents a small but significant step in the direction of greater quantification. Though the values of zero, 10, 20, and 30 on *ThisScore*, *ThatScore*, *HisScore*, and *HerScore* are still treated only as nominal identifiers, the frequency with which each value recurs is now taken into account.

The *LocalMaximum* and *LocalMinimum* Operations

If we can accept that the values of *ThisScore*, *ThatScore*, *HisScore*, and *HerScore* relate to at least an ordinal scale of measurement, we can go on to summarize each location's scores in a more quantitative manner. One way to do this would be to select whichever of each location's four scores is highest. A similar (and, in this case, probably wiser) approach might be to select whichever is lowest.

These functions can respectively be expressed as variations on the *LocalMaximum* and *LocalMinimum* operations already introduced. Here, we merely extend those operations by allowing for more than a single existing map layer as input.

The statements for these operations are specified as

NEWLAYER   = *LocalFUNCTION of FIRSTLAYER [and NEXTLAYER] etc.*

where both *FIRSTLAYER* and *NEXTLAYER* may be the titles of existing map layers or numerals representing constants.

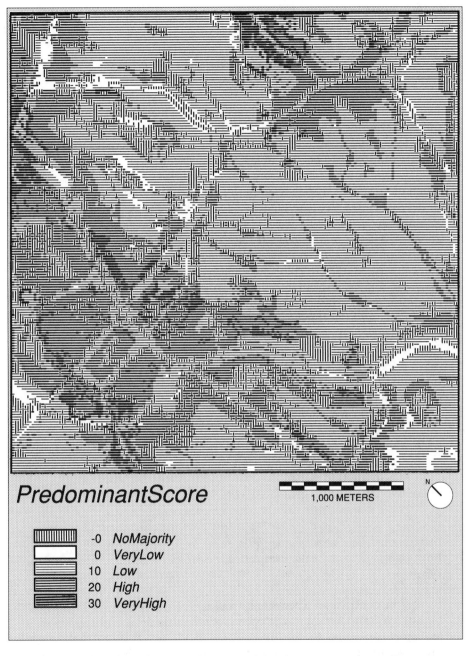

PredominantScore

1,000 METERS

N

	-0	*NoMajority*
	0	*VeryLow*
	10	*Low*
	20	*High*
	30	*VeryHigh*

Figure 4-14 A map layer created using the *LocalMajority* operation. *PredominantScore* is a layer on which each location in the Brown's Pond study area is set to a value indicating which of four land development suitability scores occurs most often at that location.

An example of *LocalMinimum* is presented in Fig. 4-15. Here, an operation specified as

LowScore = *LocalMinimum of ThisScore and ThatScore and HisScore and HerScore*

has been used to generate a layer on which each location's value indicates the most pessimistic (and therefore perhaps the most useful) of its four housing suitability assessments. Those sites that fare well on *LowScore* are likely to be safe investments.

Operations *LocalMaximum* and *LocalMinimum* can also be used in a much more specific capacity: to perform logical functions on local values in the manner of Boolean algebra. To do this requires only that a consistent pair of values be used to represent true and false conditions as zones on a set of map layers. If the true condition is represented by the greater of these two values, then *LocalMinimum* and *LocalMaximum* can respectively be used to compute what amount to cartographic *intersections* and *unions*. An intersection of true/false conditions yields a true condition as its result only if all of the input conditions (the first *and* the second *and* the third ...) are true. A union of true/false conditions yields a true condition when any of the input conditions (the first *or* the second *or* the third ...) are true.

Examples of this use of *LocalMaximum* and *LocalMinimum* are presented in Figs. 4-16 and 4-17. The *ThisAndThat* layer in Fig. 4-16 identifies those locations that are
- near Brown's Pond and
- near another pond

while the *ThisOrThat* layer in Fig. 4-17 identifies locations that are
- near Brown's Pond or
- near another pond.

These two layers were generated as follows:

ThisClose = *LocalRating of ThisFar with 1 for ... 500 with -1 for 501 ...*
ThatClose = *LocalRating of ThatFar with 1 for ... 500 with -1 for 501 ...*
ThisAndThat = *LocalMinimum of ThisClose and ThatClose*
ThisOrThat = *LocalMaximum of ThisClose and ThatClose*

In contrast to operations *LocalRating*, *LocalCombination*, *LocalVariety* , *LocalMajority*, and *LocalMinority*, operations *LocalMaximum* and *LocalMinimum* are more sensitive to the magnitude of values than to their mere presence, absence, or frequency of occurrence. As such, these are operations that begin to suggest both the power and the responsibility of increasing quantification.

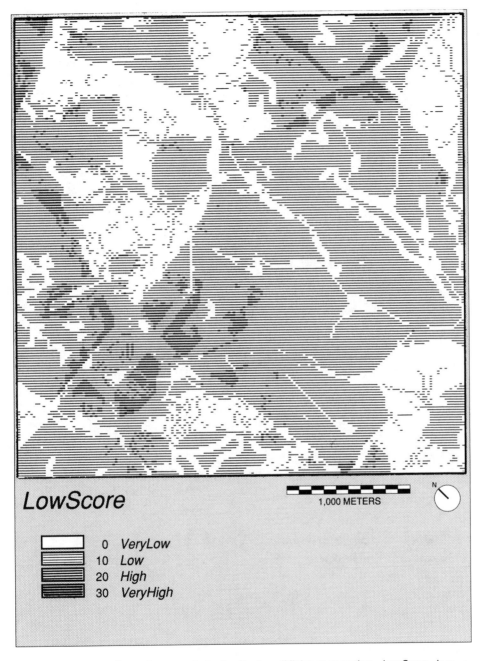

LowScore

1,000 METERS

	0	VeryLow
	10	Low
	20	High
	30	VeryHigh

Figure 4-15 A map layer created using the *LocalMinimum* operation. *LowScore* is a layer on which each location in the Brown's Pond study area is set to the lowest of four land development suitability scores assigned to that location.

Figure 4-16   A map layer created using the *LocalMinimum* operation to perform a logical intersection.   *ThisAndThat* is a layer depicting those locations within the Brown's Pond study area that are near Brown's Pond *and* near other ponds.

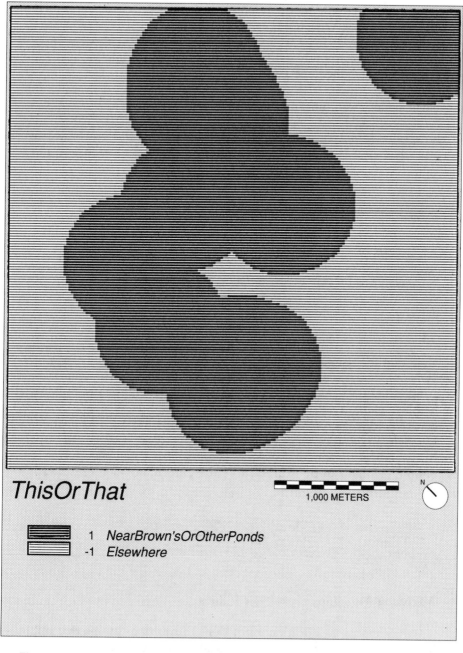

ThisOrThat

1,000 METERS

N

1 NearBrown'sOrOtherPonds
-1 Elsewhere

Figure 4-17 A map layer created using the *LocalMaximum* operation to perform a logical union. *ThisOrThat* is a layer depicting those locations within the Brown's Pond study area that are near Brown's Pond *or* near other ponds.

The *LocalSum, LocalDifference, LocalProduct, LocalRatio, LocalRoot, LocalArcTangent* **and** *LocalMean* **Operations**

To take full advantage of values relating to interval and ratio scales, we introduce one final set of local operations. Like *LocalRating*, *LocalMaximum*, and *LocalMinimum*, most are variations on operations that have already been introduced. These include versions of *LocalSum, Local-Difference, LocalProduct, LocalRatio, LocalRoot*, and *LocalArcTangent* that can now be applied to two or more existing map layers. To this group, we also add a *LocalMean* operation to compute the average of each location's values on two or more map layers.

The statement for each of these operations is specified as

NEWLAYER   = *LocalFUNCTION of FIRSTLAYER [and NEXTLAYER] etc.*

An example of *LocalMean* is presented in Fig. 4-18. Here, the values of *ThisScore, ThatScore, HisScore*, and *HerScore* at each location have been averaged by way of an operation specified as

AverageScore   = *LocalMean of ThisScore and ThatScore*
                                 and HisScore and HerScore

Note that the use of *LocalMean* here implies that the values of *ThisScore, ThatScore, HisScore*, and *HerScore* relate at least to an interval scale.

Another example of the use of these operations is presented below. Here, *LocalDifference, LocalProduct*, and *LocalRatio* are used to generate a layer on which each location's value indicates its change in cost over a period of time. Given *CostThen* and *CostNow* as layers respectively indicating the land costs at the beginning and the end of that period, change can be expressed as the percentage of each location's *CostThen* value as follows:

CostDifference   = *LocalDifference of CostNow and CostThen*
CostDifference   = *LocalProduct of CostDifference and 100*
CostDifference   = *LocalRatio of CostDifference and CostThen*

Additional Functions of Multiple Values

Among the additional functions of multiple values that might be used to characterize individual locations are exponential products, logarithms, nth highest or nth lowest values, nth most frequent or nth least frequent values, number of values above or below a specified value,

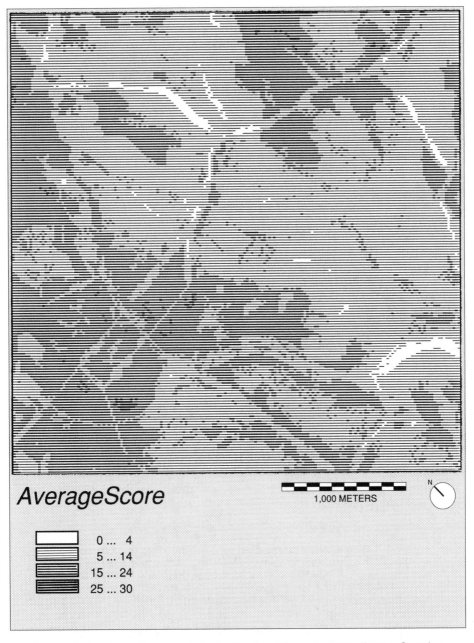

AverageScore

1,000 METERS

N

	0 ... 4
	5 ... 14
	15 ... 24
	25 ... 30

Figure 4-18 A map layer created using the *LocalMean* operation. *AverageScore* is a layer on which each location in the Brown's Pond study area is set to the average of four land development suitability scores assigned to that location. Note that each shading pattern represents not just one zone but a range of averages.

geometric means, harmonic means, quadratic means, medians, percentiles, interpercentile ranges, deviations, mean deviations, standard deviations, variances, measures of skewness or kurtosis, standard errors, coefficients of variation, z-scores, and so on. Many of these functions can be implemented by combining more primitive operations.

In Fig. 4-19, for example, is a layer on which each location in the Brown's Pond study area is set to the standard deviation of its *ThisScore*, *ThatScore*, *HisScore*, and *HerScore* values. This *StandardScore* layer was generated as follows:

AverageScore	=	*LocalMean of ThisScore and ThatScore and HisScore and HerScore*
ThisDeviation	=	*LocalDifference of ThisScore and AverageScore*
ThisDeviation	=	*LocalProduct of ThisDeviation and ThisDeviation*
ThatDeviation	=	*LocalDifference of ThatScore and AverageScore*
ThatDeviation	=	*LocalProduct of ThatDeviation and ThatDeviation*
HisDeviation	=	*LocalDifference of HisScore and AverageScore*
HisDeviation	=	*LocalProduct of HisDeviation and HisDeviation*
HerDeviation	=	*LocalDifference of HerScore and AverageScore*
HerDeviation	=	*LocalProduct of HerDeviation and HerDeviation*
Variance	=	*LocalMean of ThisDeviation and ThatDeviation and HisDeviation and HerDeviation*
StandardScore	=	*LocalRoot of MeanDeviation and 2*

4-3 QUESTIONS

This chapter has defined a set of cartographic modeling capabilities that provide for the explicit characterization of individual locations. To further review these capabilities, consider the following questions.

4-1 Given a layer of three-digit values, how could you decompose this into one new layer recording each original value's first digit, another recording its second digit, and another its third such that a value of 724 would yield a seven on the first new layer, a two on the second, and a four on the third?

4-2 "If the site is wooded, indicate the predominant forest type. Otherwise, show water bodies unless they occur on developed land, in which case the development type should appear." How would this sort of "if-then" query be expressed in cartographic modeling terms?

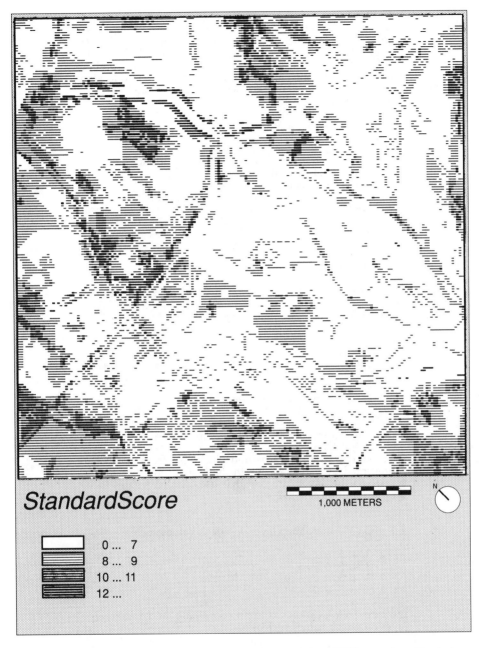

Figure 4-19 A map layer created using operations *LocalDifference*, *LocalProduct*, *LocalRoot*, and *LocalMean*. *StandardScore* is a layer on which each location in the Brown's Pond study area is set to the standard deviation of four land suitability scores. Note that each shading pattern represents not just one zone but a range of standard deviations.

4-3 *DryLand* is a layer on which values of zero and one represent wet and dry conditions, respectively. *FlatLand* is a similar layer on which zero and one represent steep and flat topography, while *OpenLand* uses the same values to distinguish forests from open fields. How would you combine these layers into one on which zero and one are used to identify all those locations that are dry and flat, or open?

4-4 The weighted-average of a set of values can be calculated by

- multiplying each value with a numerical "weight" indicating its relative importance,
- adding the resulting products, and
- dividing that sum by the sum of the absolute values of all weights assigned.

How would you weight-average *ThisScore*, *ThatScore*, *HisScore*, and *HerScore* with weights of 10, 25, 25, and 40, respectively? Why not use equivalent weights of two, five, five, and eight?

4-5 What is the (potential) difference between

LayerOneTwo	=	*LocalMean of FIRSTLAYER and SECONDLAYER*
LayerThreeFour	=	*LocalMean of THIRDLAYER and FOURTHLAYER*
NEWLAYER	=	*LocalMean of LayerOneTwo and LayerThreeFour*

and

NEWLAYER	=	*LocalMean of FIRSTLAYER and SECONDLAYER and THIRDLAYER and FOURTHLAYER*

How can this difference be mitigated?

4-6 Topographic *aspect* is the compass direction toward which a hillside descends. Given a layer on which aspect is expressed in degrees measured clockwise from north and represented by values of one through 360, how would you transform this layer into one that uses similar values to indicate uphill directions?

4-7 *EachColumn* is a map layer on which values such as

```
1 2 3 4 5 6 7 8 9
1 2 3 4 5 6 7 8 9
1 2 3 4 5 6 7 8 9
1 2 3 4 5 6 7 8 9
```

indicate each location's column coordinate, and *EachRow* is a layer on which values such as

```
1 1 1 1 1 1 1 1 1
2 2 2 2 2 2 2 2 2
3 3 3 3 3 3 3 3 3
4 4 4 4 4 4 4 4 4
```

indicate row coordinates. How could these two layers be combined to generate a layer of proximity to the location at column 11, row 73? How about a layer indicating every location's compass bearing with respect to that location?

4-8 One way to classify location-characterizing operations is in terms of the nominal, ordinal, interval, or ratio nature of the data to which they can be applied. How do *LocalArcCosine*, *LocalArcSine*, *LocalArcTangent*, *LocalCombination*, *LocalCosine*, *LocalDifference*, *LocalMajority*, *LocalMaximum*, *LocalMean*, *LocalMinimum*, *LocalMinority*, *LocalProduct*, *LocalRating*, *LocalRatio*, *LocalRoot*, *LocalSine*, *LocalSum*, *LocalTangent*, and *LocalVariety* relate to these scales of measurement?

4-9 Another way to classify these operations is in terms of the general way in which each summarizes a set of values. Some generate new values that are *typical* of the existing values associated with each location. Others generate new values that are *atypical* of the existing values they serve to represent. And still others summarize each set of values by measuring their *variation*. How do the location-characterizing operations relate to these three classes?

Chapter 5

CHARACTERIZING LOCATIONS WITHIN NEIGHBORHOODS

The operations so far described provide for the interpretation of characteristics at individual locations. They are not concerned with the relationship between one location and another and, as such, are essentially nonspatial in nature. These location-characterizing operations offer interpretive capabilities much like those of a general-purpose statistical program or data base management system.

The same is not true, however, of our second group of operations. Each of these operations computes a new value for every location as a function of its neighborhood. As indicated earlier, a *neighborhood* is any set of one or more locations that bear a specified distance and/or directional relationship to a particular location, the neighborhood *focus*. The operations in this group correspond to what are elsewhere called *convolution*, *filtering*, or *roving window* transformations.

Neighborhood-characterizing operations can be classified according to the nature of the spatial relationship between each neighborhood and its focus. In particular, a distinction can be drawn between
- operations that generate new values as a function of the existing values of locations in the immediate vicinity of each neighborhood focus, and
- operations that generate new values as a function of existing values within an extended vicinity.

5-1 FUNCTIONS OF IMMEDIATE NEIGHBORHOODS

The first group of neighborhood-characterizing operations includes those in which the neighborhoods involved are limited to locations immediately adjacent to the neighborhood focus.

As illustrated in Fig. 5-1, each location may be surrounded by as many as eight adjacent neighbors. Associated with each of these neighbors are its explicit values on various layers and a pair of implicit measures that, respectively, indicate its distance and its direction from the neighborhood focus. Any one or more of these properties may be used to characterize locations within neighborhoods. Among the operations to be introduced for this purpose are

- *FocalRating, FocalCombination, FocalVariety, FocalMajority, FocalMinority, FocalMaximum, FocalMinimum, FocalSum, FocalProduct*, and *FocalMean,*
- *FocalPercentage, FocalPercentile*, and *FocalRanking ,*
- *FocalInsularity,*
- *IncrementalLinkage* and *IncrementalLength,*
- *IncrementalPartition, IncrementalFrontage*, and *IncrementalArea*, and
- *IncrementalVolume, IncrementalGradient, IncrementalAspect*, and *IncrementalDrainage.*

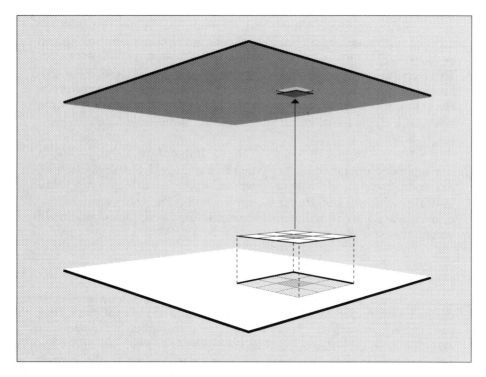

Figure 5-1   Functions of immediate neighborhoods. Operations *FocalCombination, FocalInsularity, FocalMajority, FocalMaximum , FocalMean, FocalMinimum, FocalMinority, FocalPercentage, FocalPercentile, FocalProduct, FocalRanking , FocalRating , FocalSum, FocalVariety, IncrementalArea, IncrementalAspect, IncrementalPartition, Incremental-Drainage, IncrementalFrontage, IncrementalGradient, IncrementalLength, Incremental-Linkage,* and *IncrementalVolume* can all be used to compute a new value (above) for each location as a function of its immediate neighbors on an existing layer (below).

The *FocalRating, FocalCombination, FocalVariety, FocalMajority, FocalMinority, FocalMaximum, FocalMinimum, FocalSum, FocalProduct,* **and** *FocalMean* **Operations**

Among the most fundamental of the neighborhood-characterizing operations are those that combine values within immediate neighborhoods in much the same way as values at individual locations are combined by *LocalFUNCTION* operations. Like their location-characterizing counterparts, these operations vary according to the nominal, ordinal, interval, or ratio nature of the data to which they can be applied. Also like their counterparts, they vary according to whether their purpose is to characterize the variation within a set of existing values, to summarize those values in the form of a statistic representing a typical value, or to summarize them in terms of a value that is atypical of the group.

Ratio, interval, ordinal, and even nominal values can be summarized within immediate neighborhoods by way of operations given as *FocalRating, FocalCombination, FocalVariety, FocalMajority,* and *FocalMinority* . These operations employ functions similar to those of *LocalRating, LocalCombination, LocalVariety, LocalMajority,* and *LocalMinority,* respectively. Here, however, the functions are applied not to the values of multiple layers at a single location but to the values of multiple locations on a single layer.

An example of one of these immediate neighborhood-characterizing operations is presented in Fig. 5-2. Here, *FocalCombination* has been used to generate a layer on which each location's value indicates which of the *Water* zones presented in Fig. 1-10 occur(s) atone or more loactions within its immediate neighborhood. The operation involved was specified as

ShoreType = FocalCombination of Water

If the values of locations in immediate neighborhoods are defined with respect to at least an ordinal scale of measurement, they can also be summarized by way of operations analogous to *LocalMaximum* and *LocalMinimum*. These operations are given as *FocalMaximum* and *FocalMinimum,* respectively. *FocalMaximum* sets each location to the highest (and *FocalMaximum* the lowest) of the values in its immediate neighborhood on an existing map layer.

To characterize immediate neighborhoods in terms of interval- or ratio-scale data, we can also develop analogues to other location-characterizing functions. In doing so, however, we must recognize that certain of those functions are sensitive to the sequence in which existing

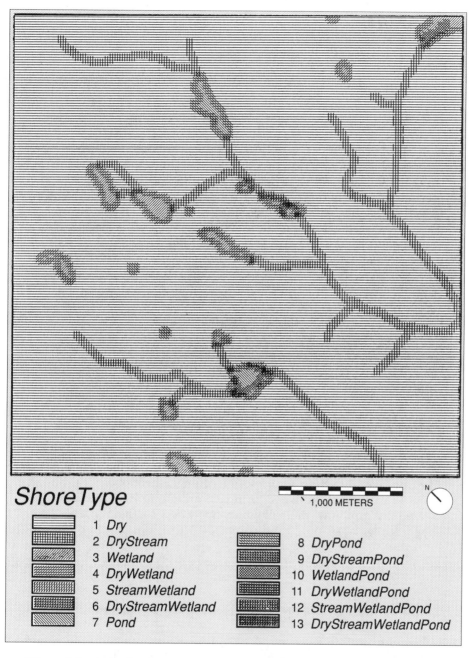

Figure 5-2 A map layer created using the *FocalCombination* operation. *ShoreType* is a layer uniquely identifying to the combination of *Water* zones that occur in the immediate vicinity of each location within the Brown's Pond study area.

values are processed. An operation specified as

TheAnswer   = *LocalDifference of ThisLayer and ThatLayer*

for example, will produce results that may be quite different from

TheAnswer   = *LocalDifference of ThatLayer and ThisLayer*

Here, processing order is defined by the sequence in which existing layers are specified. In the case of neighborhoods, there is no comparable linear order in which values should be processed. This precludes the use of functions such as subtraction, division, roots, and arc tangents. It does not affect commutative functions, however, such as those associated with operations *LocalSum*, *LocalProduct*, and *LocalMean*. The neighborhood analogues to these operations are given as *FocalSum*, *FocalProduct*, and *FocalMean*, respectively.

       The *FocalCombination, FocalVariety , FocalMajority, FocalMinority, Focal-Maximum, FocalMinimum , FocalSum, FocalProduct ,* and *FocalMean* statements are all specified as

NEWLAYER   = *FocalFUNCTION of FIRSTLAYER*

The *FocalRating* statement is also specified in this format but, in addition, may include one or more assignment phrases as follows:

NEWLAYER   = *FocalRating of FIRSTLAYER*
                     [with NEWVALUE for FIRSTZONES] etc.

       An example of *FocalRating* is presented in Fig. 5-3. Here, *Woods-Edge* is a layer depicting all of those locations where open land lies adjacent to hardwood, softwood, or mixed growth forest. This layer was created by applying the following operation to the *Vegetation* layer shown in Fig. 1-11.

WoodsEdge   = *FocalRating of Vegetation with 1 for 0 1 with 2 for 0 2 with 3 for 0 3*

The *FocalPercentage, FocalPercentile ,* and *FocalRanking* Operations

       As indicated earlier, locations within immediate neighborhoods do not exhibit the kind of one-dimensional ordering of values that is needed to define operations such as *LocalDifference, LocalRatio, LocalRoot,* or *LocalArcTangent*. Locations in immediate neighborhoods do

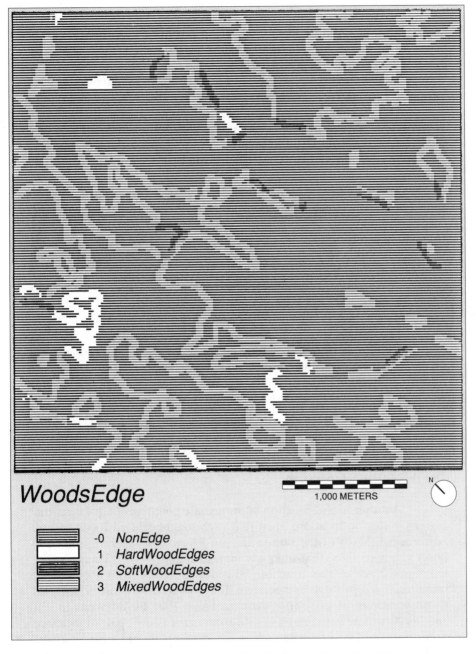

Figure 5-3 A map layer created using the *FocalRating* operation. *WoodsEdge* is a layer identifying all locations within the Brown's Pond study area where open land and particular types of forest vegetation are exclusively adjacent to one another.

exhibit a two-dimensional ordering, however, that can be used to define other neighborhood-characterizing operations. Among the most basic of these are operations that characterize each neighborhood by relating the value of its focus to those of the overall neighborhood.

FocalPercentage, for example, is an operation that determines the proportion of each neighborhood that shares the value of the neighborhood focus on a specified map layer. It then expresses that value as a percentage of total number of locations within the neighborhood and assigns it to the focal location on a new map layer. This operation can be applied to nominal- as well as ordinal-, interval-, or ratio-scale data.

For values of at least ordinal significance, a useful variation on the *FocalPercentage* function is one that indicates the percentage of each neighborhood with value(s) less than that of the focus. This is the function of operation *FocalPercentile*. *FocalRanking* is a similar operation but one that records the number of zones, rather than locations, with value(s) less than that of the focus.

The statement for each of these operations is specified as

NEWLAYER   = *FocalFUNCTION of FIRSTLAYER*

In Fig. 5-4 is an example of *FocalPercentile*. Here, *HowProminent* is a layer on which each location's value indicates the proportion of its immediate neighborhood that occurs at a lower elevation. This layer was generated from the *Altitude* layer shown in Fig. 1- 9 as follows:

HowProminent   = *FocalPercentile of Altitude*

The *FocalInsularity* Operation

Another useful function of immediate neighborhoods is one that assigns values to locations such that each neighborhood focus and all adjacent neighbors of the same value on a specified layer are ultimately set to a new value that uniquely identifies that group. To appreciate the significance of this function, consider its effect from a bird's eye rather than worm's eye perspective. If a unique new value is assigned to all locations of the same existing value that lie adjacent to one another, that new value must extend from one adjacent pair of locations to the next until it encompasses all such locations within a *conterminous* (all within a common boundary) "island." Since this new value can occur nowhere else, it will distinguish each of these insular groups of locations (sometimes called *clumps*) from every other such group.

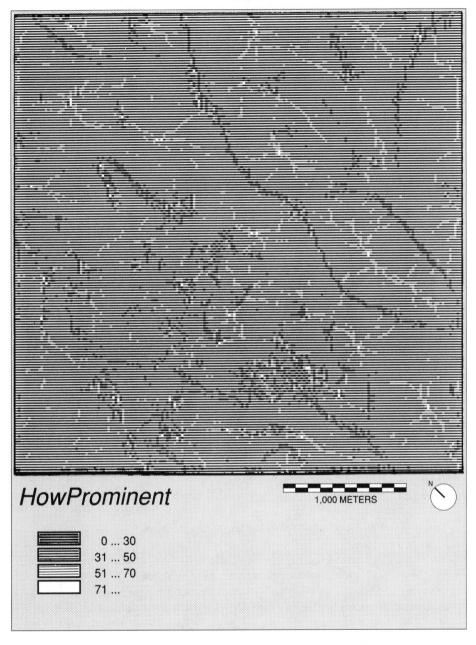

HowProminent

1,000 METERS

N

▓▓▓	0 ... 30
▨▨▨	31 ... 50
☰☰☰	51 ... 70
☐	71 ...

Figure 5-4 A map layer created using the *FocalPercentile* operation. *HowProminent* is a layer on which each location within the Brown's Pond study area is set to a percentage value indicating how much of its immediate neighborhood is at a lower altitude. Note that each shading pattern represents not just one zone but a range of percentages.

Some of these groups may be no larger than a single location, while others may encompass entire zones. Regardless of size, however, each insular group of locations will appear as a separate zone on a new map layer. In digital image processing, this is referred to as *component labeling*.

The operation embodying this function is *FocalInsularity*. Its statement is specified as

NEWLAYER = *FocalInsularity of FIRSTLAYER*

One example of *FocalInsularity* can be seen in Fig. I-1. Here, Brown's Pond was distinguished from six other ponds in the Brown's Pond study area by applying the following procedure to the *Water* layer shown in Fig. 1-10:

EveryPond	=	*LocalRating of Water with -0 for 0 ... 2*
EachPond	=	*FocalInsularity of EveryPond*
WhichPond	=	*LocalRating of EachPond with 0 for -0 with 2 for 1 ... 6 with 1 for 2*

Another example of *FocalInsularity* is shown in Fig. 5-5. Here, *EachBlock* is a layer on which each continuous portion of the Brown's Pond study area that is unbroken by major or minor roads has been set to a different value. Note that these values begin with one at the upper left corner of the study area and increase by one as each new "block" is encountered in a left-to-right sweep across the uppermost row, then the next row below, and so on. *EachBlock* was generated from the *Development* layer shown in Fig. 1-12 as follows:

EveryBlock	=	*LocalRating of Development with -1 for ... 5 with -0 for 1 2*
EachBlock	=	*FocalInsularity of EveryBlock*

The *IncrementalLinkage* and *IncrementalLength* Operations

By giving greater consideration to the spatial position of neighborhood locations in relation to neighborhood focuses, we can develop additional functions of immediate neighborhoods that begin to address the geometry of cartographic forms. As indicated earlier (in Sec. 1-9), these cartographic forms may be punctual, lineal, areal, or surficial.

Punctual configurations can generally be characterized by way of the operations already introduced. Lineal, areal, and surficial conditions, however, demand more specialized treatment.

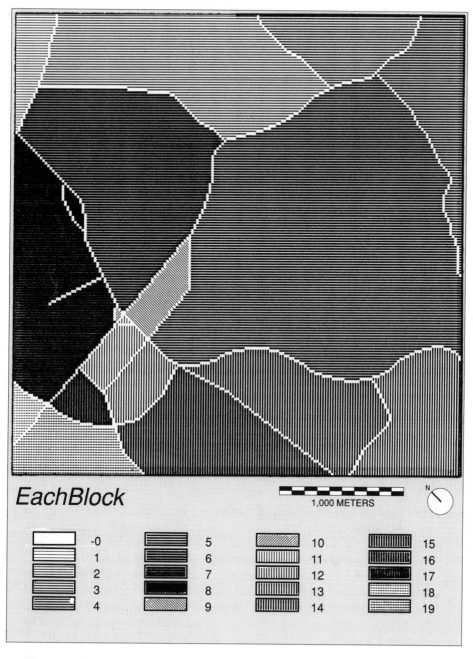

EachBlock

1,000 METERS

-0	5	10	15
1	6	11	16
2	7	12	17
3	8	13	18
4	9	14	19

Figure 5-5 A map layer created using the *FocalInsularity* operation. *EachBlock* is a layer on which each continuous expanse of land between roads within the Brown's Pond study area is set to a unique new value.

To characterize lineal shape and size within immediate neighborhoods, we introduce operations *IncrementalLinkage* and *IncrementalLength*. Each of these operations accepts input in the form of an existing layer on which all locations of value other than -0 are presumed to represent lineal conditions whose form is inferred as shown in Fig. 1-16. *IncrementalLinkage* generates a new map layer on which each location is set to a value identifying this form in terms of the 47 "crow's foot" types presented in Fig. 1-17. *IncrementalLength* reports the length of this form at each location.

The statements for these operations are specified as

NEWLAYER = *IncrementalFUNCTION of FIRSTLAYER*

The *IncrementalLength* operation can also take into account the effect of surficial conditions inferred as illustrated in Fig. 1-21. To do so, its statement should be specified as

NEWLAYER = *IncrementalLength of FIRSTLAYER on SURFACELAYER*

where *SURFACELAYER* is the title of a layer of surface elevations. In this case, each of the line segments inferred as shown in Fig. 1-17 is associated with endpoints whose vertical positions are defined by a surface like that shown in Fig. 1-21. The length of each line segment is thus increased by the secant of its angle with the cartographic plane.

The *IncrementalPartition*, *IncrementalFrontage*, and *IncrementalArea* Operations

To characterize areal shapes and sizes within immediate neighborhoods, we introduce operations *IncrementalPartition*, *IncrementalFrontage*, and *IncrementalArea*. These operations accept input in the form of an existing layer on which values of other than -0 are taken to represent areal conditions. Like *IncrementalLinkage* and *IncrementalLength*, each of these operations characterizes cartographic shape and size in increments. In this case, however, the increments involved are areal.

The *IncrementalPartition* operation indicates which of the 15 types of areal form shown in Fig. 1-19 can be inferred at the upper right corner of each location's grid square. The *IncrementalFrontage* and *IncrementalArea* operations assume that all four of each location's corners are inferred as shown in Fig. 1-19. *IncrementalFrontage* then measures the total length of the edge inferred between each location and any adjacent neighbor(s) of dissimilar value, while *IncrementalArea* measures each location's planar area.

The statements for these operations are specified as

NEWLAYER = *IncrementalFUNCTION of FIRSTLAYER*

The *IncrementalFrontage* and *IncrementalArea* operations may also be specified with a *on SURFACELAYER* phrase. As with *IncrementalLength*, this indicates that all calculations must account for the effect of surficial conditions that are inferred from *SURFACELAYER* values as shown in Fig. 1-21.

The *IncrementalVolume*, *IncrementalGradient*, *IncrementalAspect*, and *IncrementalDrainage* Operations

The shape and size of surficial conditions can also be characterized in increments. This is done by way of the *IncrementalVolume*, *IncrementalGradient*, *IncrementalAspect*, and *IncrementalDrainage* operations. The statements for these operations are specified as

NEWLAYER = *IncrementalFUNCTION [of FIRSTLAYER] on SURFACELAYER*

where *FIRSTLAYER* is the title of an existing layer whose values define areal forms as shown in Fig. 1-19, and *SURFACELAYER* is the title of an existing layer whose values define a surface as shown in Fig. 1-21.

The *IncrementalVolume* operation computes a new value for each location indicating the surficial volume associated with whatever portion of an areal condition is represented by that location on *FIRSTLAYER*. Each location's **surficial volume** is computed as shown in Fig. 5-6.

The *IncrementalGradient* and *IncrementalAspect* operations characterize the form of a surface in the immediate vicinity of each location on a layer in terms of its orientation. *IncrementalGradient* measures **surficial slope**, the angle at which a surface is inclined in relation to the horizontal cartographic plane. This is expressed in degrees such that a horizontal surface yields a slope value of zero, while near-vertical surfaces result in values approaching 90. *IncrementalAspect* measures **surficial aspect**, the compass direction toward which an inclined surface descends most rapidly. This direction is expressed in clockwise degrees from north such that new values of 90, 180, 270, and 360, respectively, correspond to slopes facing the east, south, west, and north, while locations associated with nonsloping surfaces are represented with a value of zero.

To measure the incremental slope of a surface inferred as shown in Fig. 1-21, the multiple facets of that surface at each location must first be translated into a single, representative plane. This is typically done by calculating the three-dimensional orientation of the one inclined plane that best "fits" the surficial form in the immediate vicinity of each location. *IncrementalGradient* and *IncrementalAspect* establish this best-fitting plane as illustrated in Fig. 5-7.

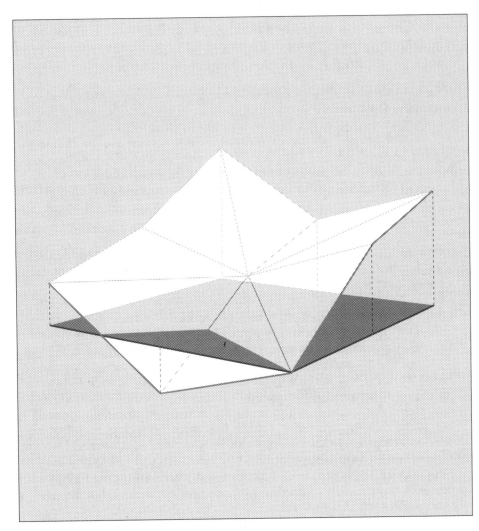

Figure 5-6 Calculation of surficial volume. The *IncrementalVolume* operation computes each location's *surficial volume* as the size of a polyhedron directly below its surficial facets (white) and above the cartographic plane (dark gray), minus the size of any polyhedron directly above those facets and below that plane.

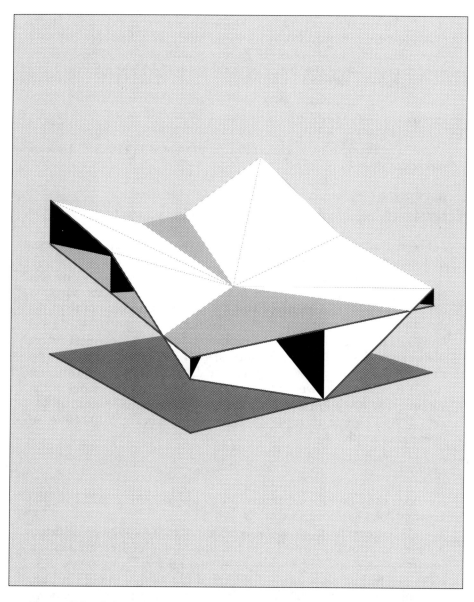

Figure 5-7 Calculation of surficial slope and aspect. Increments of *surficial slope* and *surficial aspect*, as measured by the *IncrementalGradient* and *IncrementalAspect* operations, respectively, are computed by determining that plane (light gray) whose three-dimensional orientation minimizes the sum of the squares of the vertical angles (black) between itself and a cartographic surface (white). Surficial slope is then measured as the largest angle between this inferred plane and the horizontal cartographic plane (dark gray). Surficial aspect is the horizontal direction toward which the inferred plane descends most steeply.

An example of *IncrementalGradient* is shown in Fig. 5-8. Here, each location within the Brown's Pond study area has been set to a value indicating the steepness of its topographic surface. These values were computed from the *Altitude* layer shown in Fig. 1-9 as follows:

Steepness   = *IncrementalGradient of Altitude*
Steepness   = *LocalArcTangent of Steepness*

An example of *IncrementalAspect* is presented in Fig. 5-9. Here, a procedure given as

DirectionDown   = *IncrementalAspect of Altitude*
EastSlope   = *LocalRating of DirectionDown with 0 for ... with 1 for 23 ... 157*

has been used to transform *Altitude* into a layer identifying those slopes that generally face east.

A useful variation on the *IncrementalAspect* operation is one that characterizes each location on a surface in terms of its surficial drainage. **Surficial drainage**, like surficial aspect, is a measure of the direction toward which an inclined surface descends. In this case, however, it is not the direction of steepest descent for a best-fitting plane. It is the direction(s) of **downstream** descent into each location from an adjacent neighbor. These are the directions from which streams would flow if a fluid were to be poured onto a physical model of the surface. The operation that performs this function is *IncrementalDrainage*.

An example of *IncrementalDrainage* is shown in Fig. 5-10. Here, an operation specified as

Upstream   = *IncrementalDrainage of Altitude*

has been used to trace what amount to the paths of runoff over the Brown's Pond *Altitude* surface.

To generate *IncrementalDrainage* values, each location on a surface is first characterized in terms of the direction(s) of steepest descent toward one or more of up to eight adjacent neighbors. Since any of these eight descending slopes may hold the distinction of being steepest, there may be as many as 2^8 or 256 different downstream configurations. Each of the downstream directions from one location can also be equated with an opposite **upstream** direction from the perspective of an adjacent neighbor. Thus, there are 256 different upstream configurations that can likewise occur at each location. It is this set of upstream configurations that is recorded by *IncrementalDrainage*. Each is represented by a value of from zero to 255 as indicated in Figs. 5-11 and 5-12.

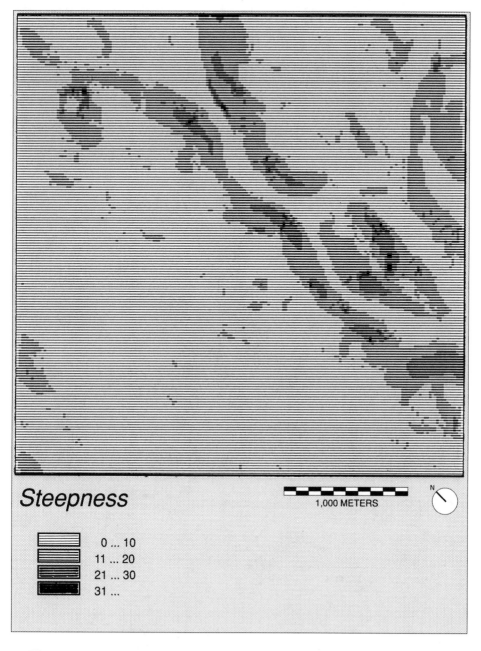

Figure 5-8 A map layer created using the *IncrementalGradient* operation. *Steepness* is a layer indicating the topographic slope at each location within the Brown's Pond study area. Here, slope is expressed as a percentage ratio of vertical to horizontal distance. Note that each shading pattern represents not just one zone but a range of percentages.

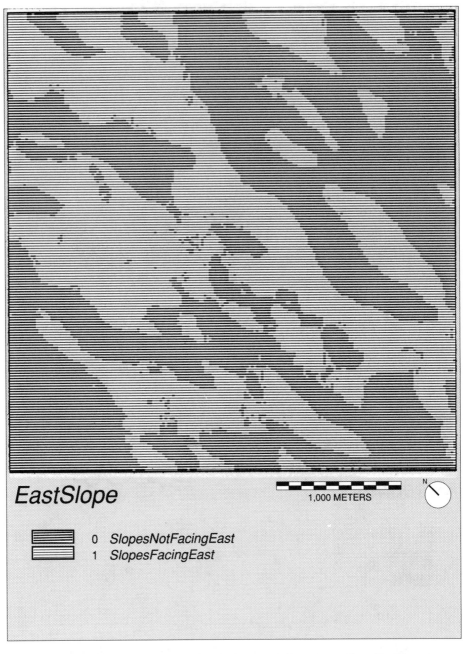

Figure 5-9 A map layer created using the *IncrementalAspect* operation. *EastSlope* is a layer indicating whether or not each location within the Brown's Pond study area is on a topographic slope facing northeast, east, or southeast.

Part II   CARTOGRAPHIC MODELING CAPABILITIES

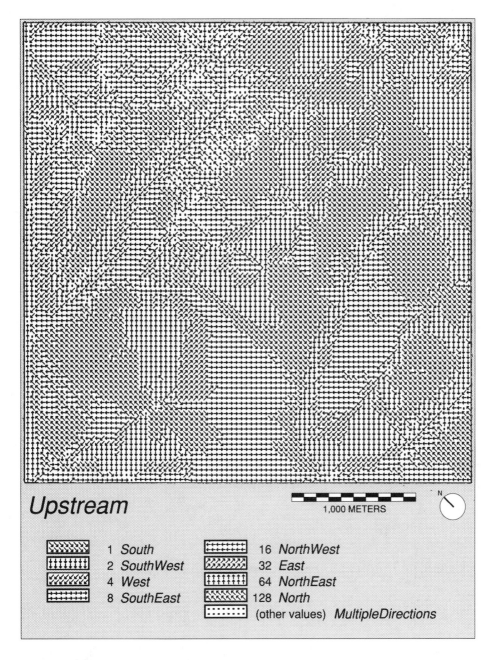

Figure 5-10 A map layer created using the *IncrementalDrainage* operation. *Upstream* is a layer indicating the direction(s) of downstream topographic descent into each location within the Brown's Pond study area. Here, only the lower right quadrant of the study area is shown.

While the *IncrementalVolume*, *IncrementalGradient*, *IncrementalAspect*, and *IncrementalDrainage* operations are most often associated with characterization of topographic surfaces, it is important to note that they can also be applied to other types of surface as well. *IncrementalGradient* and *IncrementalAspect* in particular can be used to measure the rate and/or direction of instantaneous change in any continuous quantity over space.

Figure 5-11 Calculation of surficial drainage. Increments of *surficial drainage*, as measured by the *IncrementalDrainage* operation, are determined by noting the direction(s) of steepest or downstream descent from each location on a surface toward one or more of its adjacent neighbors (arrows). These directions are then expressed in terms of the corresponding upstream directions associated with each of those neighbors (dashed lines).

Additional Functions of Immediate Neighborhoods

Locations can certainly be characterized by additional functions of their immediate neighborhoods. Most of these functions will be similar, however, to those we have introduced.

One way to anticipate additional functions is in terms of the data to which they are applied. Functions of immediate neighborhoods, like those of individual locations, will vary according to the

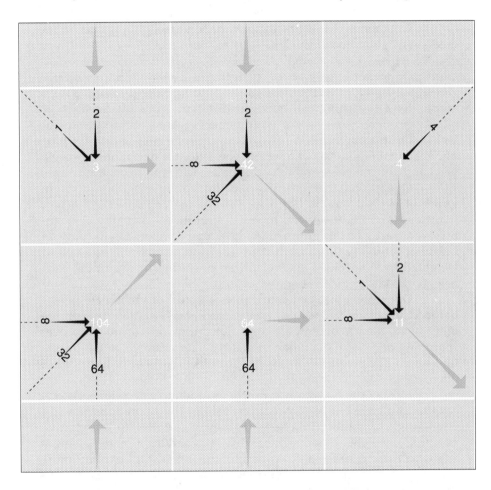

Figure 5-12 Encoding surficial drainage patterns. To encode the pattern of surficial drainage (arrows) associated with each location, the *IncrementalDrainage* operation first identifies the location's "upstream" neighbors with directional codes of 1, 2, 4, 8, 16, 32, 64, and 128 as indicated (black numbers). The sum of each location's codes (white numbers) may range from zero to 255 and will uniquely identify its particular drainage pattern.

nominal, ordinal, interval, or ratio nature of these data. Functions of immediate neighborhoods, unlike those of individual locations, may also be sensitive to the punctual, lineal, areal, or surficial nature of these data.

Another way to anticipate additional functions of immediate neighborhoods is in terms of the way in which they summarize each neighborhood's values. Much like functions of individual locations, they may do so in terms of variation, a typical case, or an atypical condition. Each of these tends to generate a distinctive spatial pattern.

Operations that characterize variation among the values in immediate neighborhoods (such as *FocalCombination*, *FocalVariety*, *FocalPercentage*, *FocalPercentile*, *FocalRanking*, *FocalInsularity*, *IncrementalLinkage*, *IncrementalLength*, *IncrementalPartition*, *IncrementalFrontage*, *IncrementalArea*, *IncrementalVolume*, *IncrementalGradient*, *IncrementalAspect*, and *IncrementalDrainage*), for example, tend to distinguish between different types of transition in quality or quantity over space. Additional functions of this type might include one that measures the range between each neighborhood's maximum and minimum values, one that computes the deviation between a neighborhood's average and the value of its focus, and so on.

Operations that characterize immediate neighborhoods in terms of typical values (such as *FocalRating*, *FocalMajority*, *FocalSum*, *FocalProduct*, and *FocalMean*) tend to soften transitions in quality or quantity over space. To this group might be added an operation to compute neighborhood averages such that each neighborhood location is weighted by a specified coefficient, an operation to determine each neighborhood's median value, and so on.

Operations that generate values atypical of immediate neighborhoods (such as *FocalMinority*, *FocalMaximum*, and *FocalMinimum*), on the other hand, tend to accentuate transitions in quality or quantity over space. An additional operation of this type, for example, might be one that selects the neighborhood value least like that which occurs most often.

Most of these additional functions of immediate neighborhoods can be implemented as procedures using the operations already introduced. For those functions that are insensitive to the relative position of neighborhood values, this is generally a matter using a *LocalFUNCTION* operation to combine the results of two or more *FocalFUNCTION* operations. Consider, for example, the following calculation of neighborhood range:

HighestValue	= *FocalMaximum of ThisLayer*
LowestValue	= *FocalMinimum of ThisLayer*
Range	= *LocalDifference of HighestValue and LowestValue*

Similarly, functions that contrast the values of each neighborhood to that of its neighborhood focus can generally be implemented by using a *LocalFUNCTION* operation to combine the new layer generated by a *FocalFUNCTION* operation with the existing layer from which it was created. For example,

MostCommonValue = FocalMajority of ThisLayer
Deviation   = LocalDifference of ThisLayer and MostCommonValue

To implement functions that are sensitive to the relative position of values within each neighborhood, we need only shift each of those values to the position of the neighborhood focus (storing the result on an intermediate layer) and then use a *LocalFUNCTION* operation to combine those eight intermediate layers with the original. The following, for example, is a procedure that computes the slope of inferred planes much like *IncrementalGradient* but which infers each best-fitting plane in terms of minimal squared deviations in neighboring values rather than angles. This procedure combines an *Altitude* layer with layers entitled *UpperLeftAltitude*, *UpperAltitude*, *UpperRightAltitude*, *RightAltitude*, *LowerRightAltitude*, *LowerAltitude*, *LowerLeftAltitude*, and *LeftAltitude*. These are layers on which each location has been set to the *Altitude* value of one of its adjacent neighbors (by way of an operation yet to be introduced in Sec. 5-2). *RESOLUTION* is a numeral representing *Altitude*'s resolution.

LeftSlopeDown   = LocalDifference of UpperLeftAltitude and LowerLeftAltitude
RightSlopeDown = LocalDifference of UpperRightAltitude and LowerRightAltitude
MidSlopeDown   = LocalDifference of UpperAltitude and LowerAltitude
MidSlopeDown   = LocalProduct of MidSlopeDown and 6
SlopeDown   = LocalSum of LeftSlopeDown and RightSlopeDown
                  and MidSlopeDown
SlopeDown   = LocalRatio of SlopeDown and 4 and RESOLUTION
SlopeDown   = LocalProduct of SlopeDown and SlopeDown
UpperSlopeLeft = LocalDifference of UpperRightAltitude and UpperLeftAltitude
LowerSlopeLeft = LocalDifference of LowerRightAltitude and LowerLeftAltitude
MidSlopeLeft   = LocalDifference of RightAltitude and LeftAltitude
MidSlopeLeft   = LocalProduct of MidSlopeLeft and 6
SlopeLeft   = LocalSum of UpperSlopeLeft and LowerSlopeLeft
                  and MidSlopeLeft
SlopeLeft   = LocalRatio of SlopeLeft and 4 and RESOLUTION
SlopeLeft   = LocalProduct of SlopeLeft and SlopeLeft
MeanSlope   = LocalSum of SlopeDown and SlopeLeft
MeanSlope   = LocalRoot of MeanSlope and 2

5-2 FUNCTIONS OF EXTENDED NEIGHBORHOODS

The second group of neighborhood-characterizing operations differs from the first in that neighborhoods may now be defined more generally. As illustrated in Fig. 5-13, a neighborhood may now include any set of locations at specified distances and/or directions with respect to its neighborhood focus. Among these operations are

- *FocalFUNCTION at DISTANCE* and *by DIRECTION*,
- *FocalProximity, FocalBearing,* and *FocalNeighbor,*
- *FocalGravitation,*
- *FocalFUNCTION radiating,*
- *FocalFUNCTION spreading,*
- *FocalFUNCTION spreading on,*
- *FocalFUNCTION spreading through,* and
- *FocalFUNCTION spreading in.*

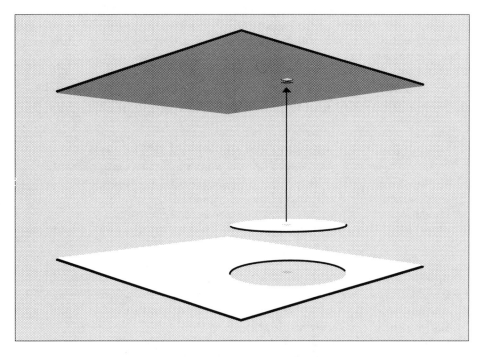

Figure 5-13 Functions of extended neighborhoods. Operations *FocalBearing, FocalCombination, FocalGravitation, FocalInsularity, FocalMajority, FocalMaximum, FocalMean, FocalMinimum, FocalMinority, FocalNeighbor, FocalPercentage, FocalPercentile, FocalProduct, FocalProximity, FocalRanking, FocalRating, FocalSum,* and *FocalVariety* can all be used to compute a new value (above) for each location as a function of those neighbors on an existing map layer (below) that lie at specified distances and/or directions.

The *FocalFUNCTION at DISTANCE* and *by DIRECTION* Operations

Operations *FocalCombination, FocalInsularity, FocalMajority, FocalMaximum, FocalMean, FocalMinimum, FocalMinority, FocalPercentage, FocalPercentile, FocalProduct, FocalRanking, FocalRating, FocalSum,* and *FocalVariety* can all be generalized by extending the size of the neighborhoods from which they compute new values. The statements for extended-neighborhood versions of these operations are specified as

NEWLAYER   = *FocalFUNCTION of FIRSTLAYER at DISTANCE*

where *DISTANCE* is one or more numerals specifying the neighborhood radius or radii in units corresponding to those of *FIRSTLAYER*'s resolution. Each neighborhood is then defined as the set of all locations whose distance from a particular focus is less than or equal to that radius.

One example of this is presented in Fig. 5-14. The *HomeDensity* layer shown here was generated by applying a procedure given as

Housing   = *LocalRating of Development with 0 for ... with 1 to 3*
HomeDensity   = *FocalSum of Housing at ... 200*

to the *Development* layer shown in Fig. 1-12.

Another example is presented in Fig. 5-15. Here, an operation specified as

SmoothVegetation   = *FocalMajority of Vegetation at ... 100*

has been applied to the *Vegetation* layer shown in Fig. 1-11 to generate a *SmoothVegetation* layer on which each location's value indicates the predominant type of forest vegetation in its vicinity. Note that the resulting cartographic pattern is a "smoothed" version of the original.

In Fig. 5-16 is a similar example of cartographic smoothing. In this case, however, the cartographic forms involved are surficial rather than areal. *SmoothAltitude* is a layer on which each location has been set to an average of neighboring topographic elevations. It was created by repeatedly applying the following operation to a version of *SmoothAltitude* initially identical to the Brown's Pond *Altitude* layer.

SmoothAltitude   = *FocalMean of SmoothAltitude at ... 1000*

As illustrated in Fig. 5-17, the result is a layer on which basins and valleys have been "filled up" (by being averaged with higher elevations nearby), while peaks and ridges have been "worn down."

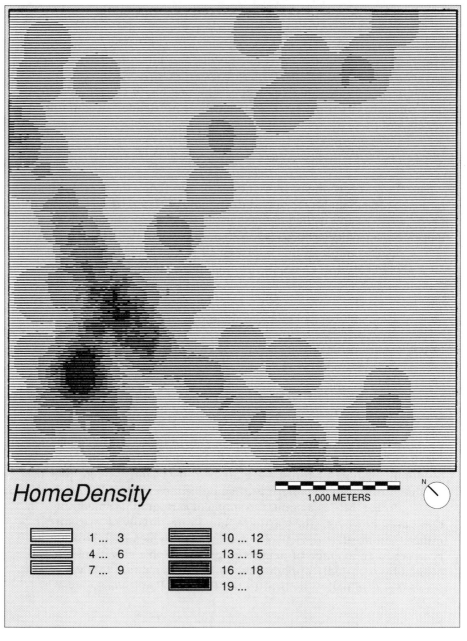

Figure 5-14 A map layer created using the *FocalSum* operation. *HomeDensity* is a layer indicating the number of homes within 200 meters of each location in the Brown's Pond study area. Note that each shading pattern represents not just one zone but a range of housing densities.

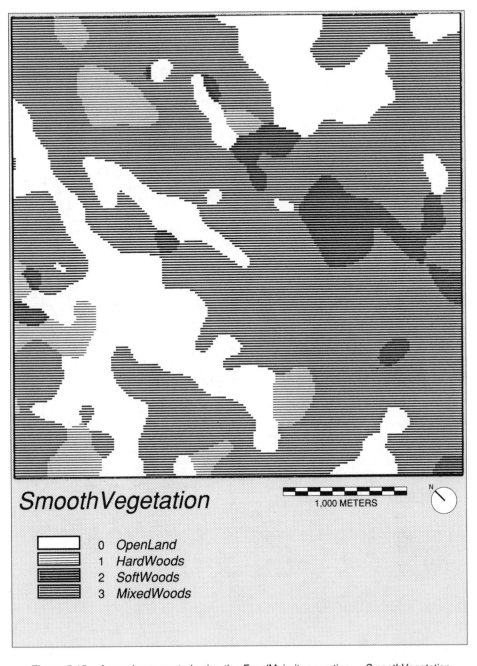

Figure 5-15 A map layer created using the *FocalMajority* operation. *Smooth Vegetation* is a layer on which each location within the Brown's Pond study area is set to the value of whatever *Vegetation* zone occurs most frequently within a radius of 100 meters.

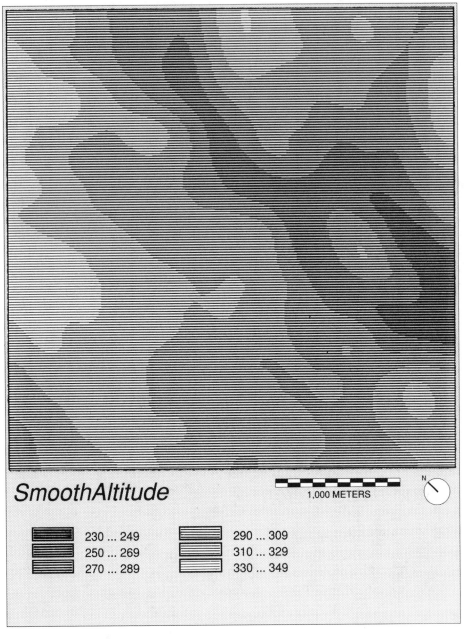

SmoothAltitude

1,000 METERS

N

230 ... 249
250 ... 269
270 ... 289

290 ... 309
310 ... 329
330 ... 349

Figure 5-16 A map layer created using the *FocalMean* operation. *SmoothAltitude* is a layer on which each location within the Brown's Pond study area is set to an average of the topographic elevation values (in meters) within its vicinity. Note that each shading pattern represents not just one zone but a range of elevations.

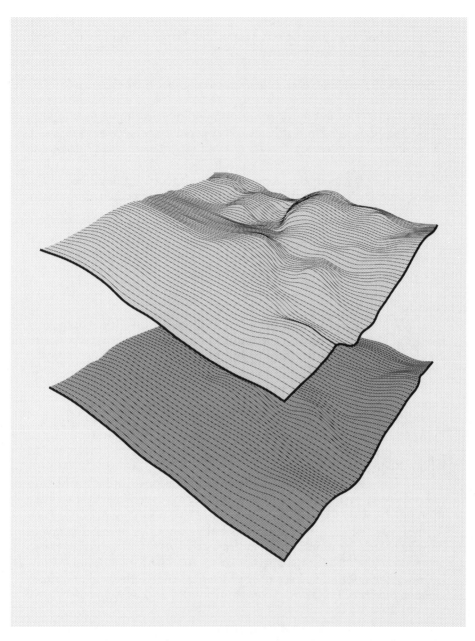

Figure 5-17 Surficial smoothing. The smoothing effect of operation *FocalMean* can be seen in perspective views of two surfaces. Above is the surface of a layer generated by repeatedly applying *FocalMean* to the Brown's Pond *Altitude* layer shown in Fig. 1-20. Below is a similar view of the *SmoothAltitude* layer shown in Fig. 5-16, the result of additional smoothing with operation *FocalMean*.

FocalFUNCTION operations can also be generalized by restricting each neighborhood to locations at specified bearings from its neighborhood focus. These are specified as

NEWLAYER = FocalFUNCTION of FIRSTLAYER by DIRECTION

where *DIRECTION* is one or more numerals indicating direction(s) in clockwise degrees from north. It is this capability that would be used, for example, to generate the *UpperLeftAltitude*, *UpperAltitude*, *UpperRightAltitude*, *RightAltitude*, *LowerRightAltitude*, *LowerAltitude*, *LowerLeftAltitude*, and *LeftAltitude* layers in the slope-measuring procedure described at the end of Sec. 5-1. The operations involved would be specified as

UpperLeftAltitude	*= FocalMaximum of Altitude at 1 ... 2 by 360*
UpperAltitude	*= FocalMaximum of Altitude at 1 by 45*
UpperRightAltitude	*= FocalMaximum of Altitude at 1 ... 2 by 90*
RightAltitude	*= FocalMaximum of Altitude at 1 by 135*

and so on.

Another example of the *to DISTANCE* and *by DIRECTION* options of *FocalFUNCTION* operations is presented in Fig. 5-18. The *SouthEastDevelopment* layer shown here was created by applying a procedure given as

SouthEastDevelopment	*= FocalMaximum of Development at ... 500 by 125 ... 145*

to the *Development* layer shown in Fig. 1-12. *SouthEastDevelopment* indicates, for every location within the Brown's Pond study area, the *Development* zone of greatest value that occurs within 500 meters and within 10 degrees of southeast.

The generality of *FocalFUNCTION* operations can be further enhanced by allowing *DISTANCE* and *DIRECTION* specifications to vary from one location to another. This is done by specifying either or both of these parameters not as numerals but as the titles of existing layers on which each location's value indicates the *DISTANCE* or *DIRECTION* to be used in establishing that particular location's neighborhood. To illustrate, consider the effect of a procedure that reflects a mountain climber's view of the *Altitude* layer shown in Fig. 1-9 as follows:

Bearing	*= IncrementalAspect of Altitude*
LeftBearing	*= LocalSum of Bearing and 170*
RightBearing	*= LocalSum of Bearing and 190*
ThingsToCome	*= FocalMaximum of Altitude*
	at ... 200 by LeftBearing by RightBearing

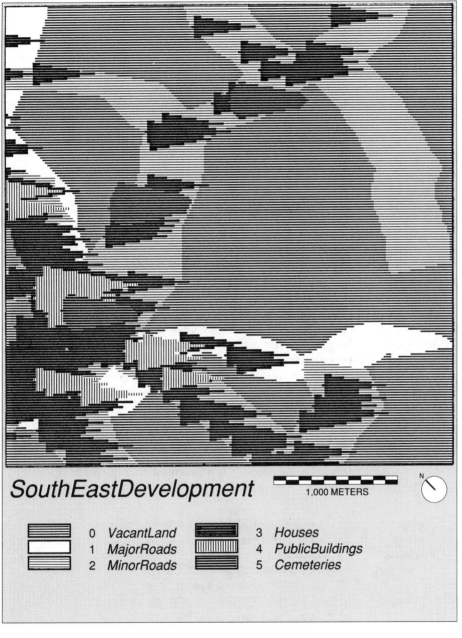

Figure 5-18 A map layer created using the *FocalMaximum at DISTANCE by DIRECTION* operation. *SouthEastDevelopment* is a layer on which each location within the Brown's Pond study area is set to a value indicating the land development type of greatest value within its southeasterly vicinity.

The *FocalProximity, FocalBearing,* and *FocalNeighbor* Operations

Just as *IncrementalFUNCTION* operations are sensitive to the relative positions as well as the values of locations in immediate neighborhoods, functions of extended neighborhoods can also be developed by considering the distance and/or the direction of each location in a neighborhood with respect to its neighborhood focus. This gives rise to three more *FocalFUNCTION* operations given as *FocalProximity, FocalBearing,* and *FocalNeighbor.*

The statements for these operations are specified as

```
NEWLAYER        = FocalFUNCTION of FIRSTLAYER
                  [at DISTANCE] [by DIRECTION]
```

FocalProximity calculates the planar distance between each location and the nearest of any locations within a neighborhood radius of *DISTANCE* that have *FIRSTLAYER* values of other than -0. This group of locations to which distance is measured is referred to as the operation's *target.* If there are no target locations within a particular neighborhood, a new value equal to *DISTANCE* is assigned. The function of *FocalProximity* is often referred to as *buffer* generation in vector-based systems and *searching* in a raster context. It is probably the most common neighborhood-processing function in general use.

Two examples of *FocalProximity* have already been presented. In Figs. I-2 and I-3 are layers that were generated by using this operation to measure distance to selected ponds.

The *FocalBearing* operation is much like *FocalProximity.* Instead of measuring distance to the nearest of a specified set of locations, however, this operation measures direction.

An example of *FocalBearing* is presented in Fig. 5-19. Here, a procedure specified as

```
PondBearing     = LocalRating of WhichPond with -0 for 0
PondBearing     = FocalBearing of PondBearing at ...
PondBearing     = LocalRating of PondBearing
                    with 1 for  338 ... 360 for 1 ... 22 with 2 for 23 ... 67
                    with 3 for  68 ... 110  with 4 for 111 ... 157
                    with 5 for 158 ... 202 with 6 for 203 ... 247
                    with 7 for 248 ... 290 with 8 for 291 ... 337
```

has been applied to the *WhichPond* layer shown in Fig. I-1 to generate a new layer on which each location's value indicates the bearing of nearest pond in one of the eight major compass directions.

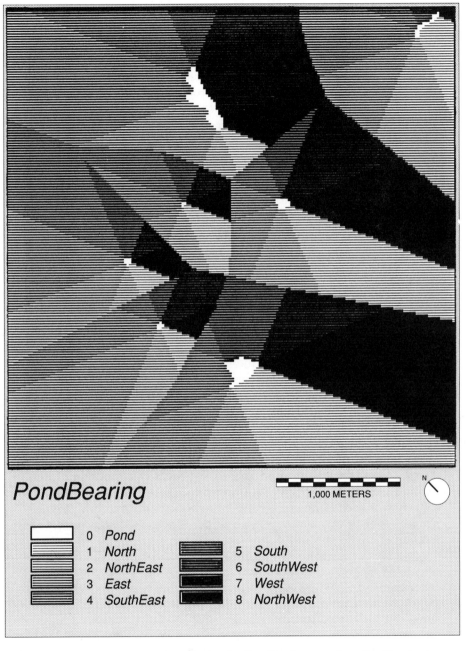

PondBearing

1,000 METERS

N

0	Pond		
1	North	5	South
2	NorthEast	6	SouthWest
3	East	7	West
4	SouthEast	8	NorthWest

Figure 5-19 A map layer created using the *FocalBearing* operation. *PondBearing* is a layer indicating the direction to the nearest pond from each location within the Brown's Pond study area.

The *FocalNeighbor* operation is also similar to *FocalProximity* and *FocalBearing*. In this case, however, it is not the proximity or the bearing of a nearest target location that is assigned to each neighborhood focus. It is the value of that nearest neighbor.

In Fig. 5-20 is an example of the result. The *NearestPond* layer shown here was created by applying a procedure specified as

EveryPond	=	*LocalRating of WhichPond with -0 for 0*
EachPond	=	*FocalInsularity of EveryPond*
NearestPond	=	*FocalNeighbor of EachPond at ...*
NearestPond	=	*LocalRating of WhichPond*
		with 0 for 1 2 with NearestPond for 0

to the *WhichPond* layer shown in Fig. I-1. For each pond, it identifies all locations from which that pond is the nearest. These sets of locations are variously referred to *Dirichlet regions*, *proximal zones*, *Theissen polygons*, *Voronoi diagrams*, and *Wigner-Seitz cells*.

The *FocalGravitation* Operation

The *FocalNeighbor* operation performs what amounts to a form of *cartographic interpolation*. To interpolate is to estimate an unknown value from a set of one or more known values. In the case of *FocalNeighbor*, this is done by setting each location of null value to the nearest of a specified set of target values.

Cartographic interpolation more typically involves estimates generated by averaging sets of two or more target values such that those nearer to a given location of unknown value exert a greater influence. Among the most widely used and most generally applicable of these methods are those that average target location values such that the influence of each is much like the influence of gravity. It is inversely proportional to the square of each target location's distance from the location of unknown value.

This is the function of operation *FocalGravitation*. Given an existing layer on which target locations are distinguished by values of other than -0, *FocalGravitation* will generate a new layer on which each target location retains its initial value, while every other location assumes a value computed by

- multiplying each target value within its neighborhood by the reciprocal of that target's squared proximity,
- adding those products, and
- dividing that sum by the sum of those reciprocals.

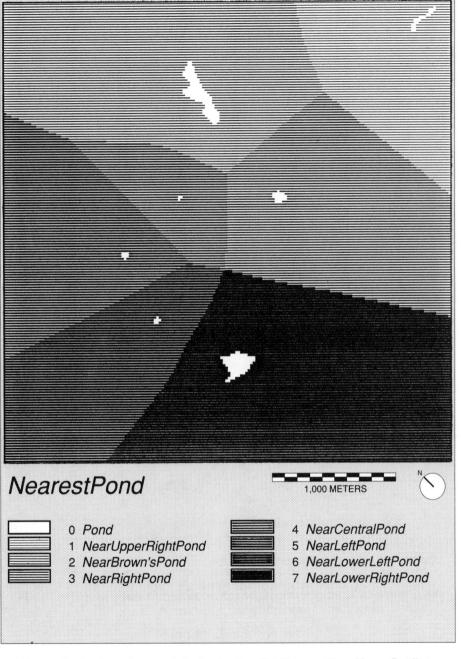

NearestPond

1,000 METERS

N

0 *Pond*	4 *NearCentralPond*
1 *NearUpperRightPond*	5 *NearLeftPond*
2 *NearBrown'sPond*	6 *NearLowerLeftPond*
3 *NearRightPond*	7 *NearLowerRightPond*

Figure 5-20   A map layer created using the *FocalNeighbor* operation. *NearestPond* is a layer indicating which pond is closest to each location within the Brown's Pond study area .

These calculations are illustrated in Fig. 5-21, and an example of *FocalGravitation* is presented in Fig. 5-22. The *InferredAltitude* layer shown here was generated by noting the value of the Brown's Pond *Altitude* layer for each location (and only those locations) occupied by a house on the Brown's Pond *Development* layer. These locations were then treated as targets in an attempt to re-create the *Altitude* surface. The procedure involved was as follows:

TargetLayer	= *LocalRating of Development with -0 for ... with Altitude for 3*
InferredAltitude	= *FocalGravitation of TargetLayer at ...*
InferredAltitude	= *LocalRating of TargetLayer with InferredAltitude for -0*
	with 0 for 0 ...

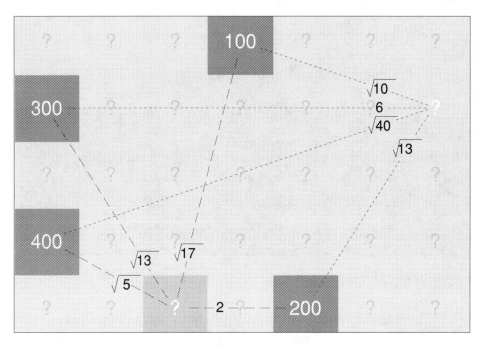

Figure 5-21 Gravitational interpolation. The *FocalGravitation* operation computes a value for each location of unknown value (question mark) by averaging the known values (white numbers) of surrounding locations (dark gray) such that each is weighted by the inverse of its distance (black number) squared. The value of the darker location in the right column, for example, would be computed as

$$((100 / 10) + (200/13) + (300/36) + (400/40)) / ((1/10) + (1/13) + (1/36) + (1/40)) = 149$$

while that of the darker location in the lower row would be computed as

$$((100 / 17) + (200/4) + (300/13) + (400/5)) / ((1/17) + (1/4) + (1/13) + (1/5)) = 308$$

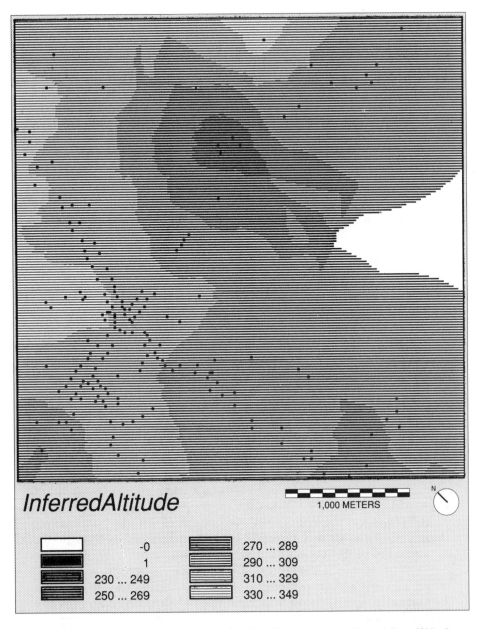

Figure 5-22 A map layer created using the *FocalGravitation* operation. *InferredAltitude* is a layer on which each location within the Brown's Pond study area is set to a new value calculated by inverse-square-distance-weight averaging the altitude of houses. Note that each shading pattern represents not just one zone but a range of topographic elevations. Note, too, how *InferredAltitude* values conform to those of *Altitude* best in the vicinity of target locations.

The *FocalFUNCTION radiating* Operation

Another useful variation on the neighborhood-characterizing capabilities of *FocalFUNCTION* operations can be developed by defining neighborhoods such that each is limited to those locations that are not only at specified distances and/or directions from the neighborhood focus but also within visual contact. To establish this visual contact, a location must be situated such that there exists an unobstructed line of sight between a designated, vertical position associated with that location and a similar position associated with the neighborhood focus. This relationship is illustrated in Fig. 5-23.

An expanded version of the *FocalFUNCTION* operation that characterizes line-of-sight neighborhoods in this manner is given as

NEWLAYER   = *FocalFUNCTION of FIRSTLAYER*
                          [at DISTANCE] etc. [by DIRECTION] etc.
                          radiating [on SURFACELAYER]
                          [from TRANSMISSIONLAYER]
                          [through OBSTRUCTIONLAYER]
                          [to RECEPTIONLAYER]

Here, the *on SURFACELAYER* phrase specifies an existing layer of values indicating position in the vertical dimension perpendicular to the cartographic plane. This is typically a layer indicating topographic elevation.

The *from TRANSMISSIONLAYER* phrase specifies a layer of values that are added to those of *SURFACELAYER* in order to establish the vertical position of each neighborhood focus. As suggested by the light bulb in Fig. 5-23, this should be envisioned as a layer indicating the height at which a point at the center of each focus's grid square is transmitting some sort of radiation. Whether these are the heights at which focuses can "see" or "be seen" will depend on the application. *TRANSMISSIONLAYER* will typically be a layer of eye-level heights or the heights of visible objects such as buildings.

The *through OBSTRUCTIONLAYER* phrase specifies a layer of values that are added to those of *SURFACELAYER* in order to establish the vertical positions below which no lines of sight can pass over any location's grid square. This is typically a layer of building or vegetation height.

And *to RECEPTIONLAYER* gives the title of a layer whose values are added to those of *SURFACELAYER* in order to establish the vertical positions of all nonfocal locations within each neighborhood. This layer indicates the height at or below which a point at the center of

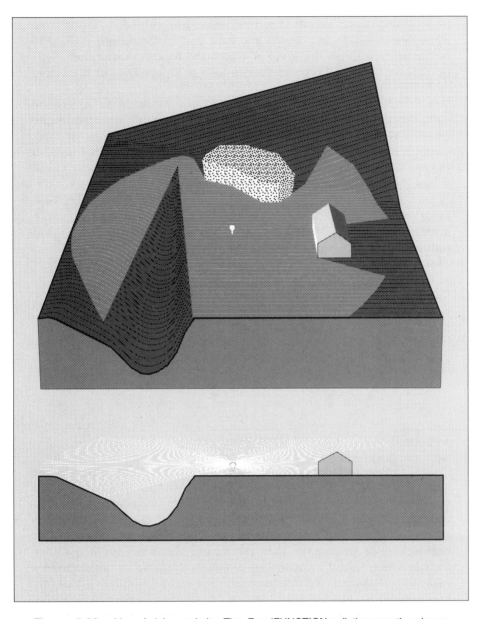

Figure 5-23 Line-of-sight proximity. The *FocalFUNCTION radiating* operation characterizes neighborhoods whose radii are measured in terms of distance and visibility from each neighborhood focus. Here, a typical situation is shown in perspective (above) and in cross-section (below). Visible locations (lightened) are those to which direct lines of sight can be constructed from a specified height above the neighborhood focus (light bulb) without being obstructed by intervening topography (gray surface) or site conditions (trees and building).

each grid square is prepared to receive the transmitted radiation from a neighborhood focus. As in the case of *TRANSMISSIONLAYER*, this height will establish the level at which the location can either "see" or "be seen" and will typically correspond to an eye level or the height of a visible object.

SURFACELAYER, *TRANSMISSIONLAYER*, *OBSTRUCTIONLAYER*, and *RECEPTIONLAYER* values all relate to units of distance corresponding to those of *FIRSTLAYER*'s resolution. Note that when any one of these layers is omitted, a layer of zeroes is assumed. Note, too, that none of the conditions represented by *TRANSMISSIONLAYER* or *RECEPTIONLAYER* values have any effect on visual obstructions.

An example of *FocalMaximum radiating* is illustrated in Fig. 5-24. Here, a layer entitled *SteepleView* has been created from the *Altitude* layer shown in Fig. 1-9 and the *Development* layer shown in Fig. 1-12 as follows:

Steeple	= *LocalRating of Development with -0 for 0 1 2 3 5*
Steeple	= *FocalInsularity of Steeple*
Steeple	= *LocalRating of Steeple with 0 for ... with 1 for 13*
SteepleView	= *FocalMaximum of Steeple at ...*
	radiating on Altitude
SteepleView	= *LocalRating of Development and SteepleView*
	with 1 for 1 ... 5 on 1 with 2 for 0 on 1
	with 3 for 1 ... 5 on 0 with 4 for 0 on 0

The *FocalFUNCTION spreading* Operation

In moving from the conception of proximity as measured by the original version of the *FocalProximity* operation to that which is affected by lines of sight, we have tacitly introduced an important notion that proximity is somehow related to motion. In the case of *FocalProximity radiating*, it is the implied transmission of light rays and the selective obstruction of these rays that establishes the basis for a distance metric. Other useful measures of proximity can also be developed by regarding distance as a consequence of motion and simulating that motion in order to calculate this distance.

In measuring the proximity of two locations in the cartographic plane, we usually make two assumptions. The first of these is an assumption that, in referring to distance between the two locations, we mean the minimum distance possible. The second is an assumption that this minimum distance between two locations is equal to the length of a straight line between them.

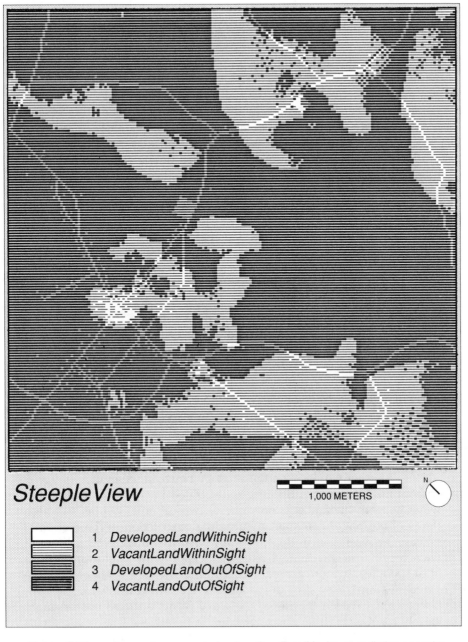

SteepleView

1,000 METERS

N

	1	*DevelopedLandWithinSight*
2	*VacantLandWithinSight*
3	*DevelopedLandOutOfSight*
4	*VacantLandOutOfSight*

Figure 5-24   A map layer created using the *FocalMaximum radiating* operation. *SteepleView* is a layer on which each location within the Brown's Pond study area is set to a new value indicating whether or not it is developed and whether or not its line of sight to the Petersham town hall steeple is blocked by intervening topography.

To measure distance as a consequence of motion, it is important that the first of these two assumptions be adopted and well understood. It is equally important, however, that the second assumption be put aside. When the distance between two locations is to be expressed in terms of measures such as the quickest time or the lowest cost associated with motion between them, the path of that motion may or may not conform to "a straight line on the map." Distance in this case is measured with respect to a *nonEuclidean* frame of reference.

One way to measure this sort of distance-as-a-consequence-of-motion is to relate that motion to the motion of waves on the surface of a body of water. To illustrate this, consider the pattern of waves that might be formed in the pool of water beneath a faucet that is dripping at a constant frequency. The resulting pattern of wave fronts at any given instant in time will form a series of concentric and evenly spaced rings around the source of those drips. Since each wave will have traveled over the shortest possible path from this source to any given position along each of these rings, the rings themselves can be equated with measures of distance to the source. This important relationship is illustrated in Fig. 5-25.

Now consider the situation shown in Fig. 5-26. Here, an obstacle has been placed in the pool of water at considerable distance from the faucet. This obstacle acts as a barrier to the motion of the waves and, as a result, they bend as they move around it. The phenomenon is known as *diffraction*. It yields a pattern of wave fronts that can no longer be equated with zones of distance "as the crow flies." They can be equated, however, with zones of distance "as the crow walks" around that barrier. And since all waves will still have traveled over shortest possible (albeit no longer radial) paths, they can still be equated with measures of proximity to the wave source.

Another variation on this analogy between the motion of waves and the measurement of distance is illustrated in Fig. 5-27. The pool of water beneath the faucet is now one of two different depths. Though the pool no longer contains any absolute obstruction to the motion of waves, this variation in depth creates what amounts to two different wave media. As waves move from the deeper to the shallower portions of the pool (or, in more general terms, from a less dense to a more dense wave medium), they slow down and are bent inward toward the shallower area. Within that area, the wave fronts are also spaced more closely together. These effects are due to *refraction*. They yield a pattern of wave fronts that cannot be equated with zones of distance either as the crow flies or walks around barriers. It is a pattern, however, that can be equated with zones of distance as that crow walks through different areas of what amount to varying **travel costs**.

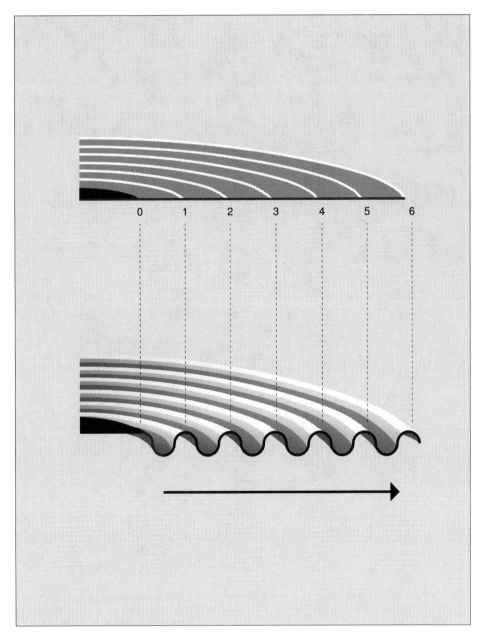

Figure 5-25 Relationships between uniform wave patterns and Euclidean distance. The instantaneous positions of wave fronts (below) moving from a wave source (black) vibrating at a constant frequency in a medium of uniform density will form a pattern of concentric circles that can be equated with zones of Euclidean distance from that source (above). In the typical case shown, the direction of wave motion is indicated by the arrow.

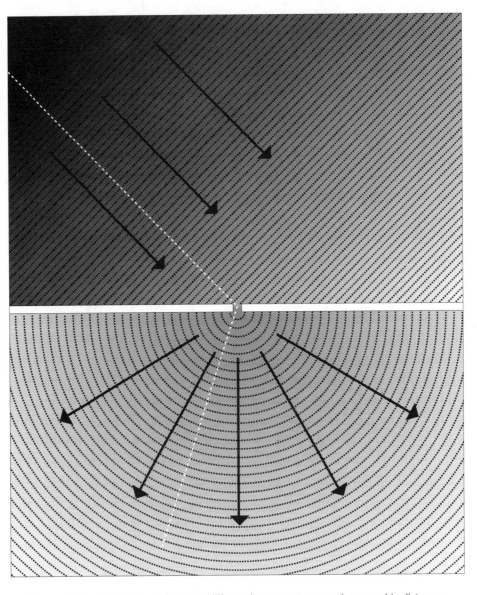

Figure 5-26 Relationships between diffracted wave patterns and geographic distance. The instantaneous positions of wave fronts moving through a medium of uniform density will form a pattern that can be equated with zones of travel-cost distance from the wave source.   In the typical case shown, the direction of wave motion is indicated by arrows, and wave fronts are drawn as black dotted lines. As waves move toward the lower right, some pass by an obstruction (white) and are diffracted as a result. This pattern of wave fronts corresponds to the pattern of travel-cost distance zones that would result if that obstruction were to act as an absolute barrier to travel. The path of wave motion drawn as a dashed white line represents a shortest path under these conditions.

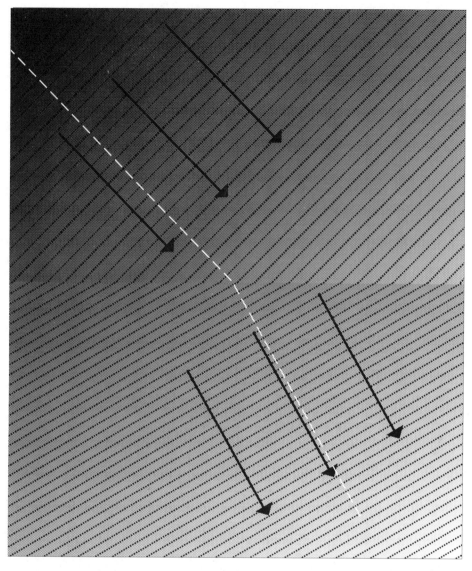

Figure 5-27 Relationships between refracted wave patterns and geographic distance. The instantaneous positions of wave fronts moving through media of varying density will form a pattern that can be equated with zones of travel-cost distance from the wave source. In the typical case shown, the direction of wave motion is indicated by arrows, and wave fronts are drawn as dotted black lines. As waves move toward the lower right, some pass into a medium of higher density and are refracted as a result. This pattern of wave fronts corresponds to the pattern of distance zones that would result if higher-density medium were to be associated with what amounts to a greater map layer resolution in units of incremental travel cost. The path of wave motion drawn as a dashed white line represents a shortest path under these conditions.

In attempting to simulate the transmission of waves over cartographic space as we have defined it, we must contend with the fact that this space is not continuous. It is made up of discrete locations, and any motion through it must therefore proceed in discrete increments. A reasonable way to do this is to propagate waves over lineal paths that extend from each location to its adjacent neighbors. This effectively treats the cartographic plane as a finite *network*. It is a network in which locations are represented as *nodes*, and increments of distance between adjacent locations are represented as weighted *links*. Such a network is illustrated in Fig. 5-28.

To generate zones of distance around a designated set of locations that are represented in this manner, we start from those locations and proceed outward in "waves" by repeatedly moving from location to adjacent location. As each new link is traversed to reach a new location, the distance of that location is calculated by adding the length of that link to the distance of the adjacent location from which it extends. If this is done such that each location is reached by the shortest path possible, resulting distances will be as shown in Fig. 5-28.

A version of the *FocalFUNCTION* operation that measures distance in this manner is given as

NEWLAYER = FocalFUNCTION of FIRSTLAYER spreading at DISTANCE

This operation differs from the earlier versions of *FocalFUNCTION* in that neighborhood radii are now measured not in terms of the length of a straight line connecting the centers of two grid squares but, rather, in terms of the length of a lineal path made up of increments.

As such, distance measurements suffer from the kind of imprecision that is illustrated in Fig. 1-23. This imprecision can also be seen in the octagonal (rather than circular) patterns that are apparent in Fig. 5-28. While this new version of the *FocalFUNCTION* operation is therefore not as effective as the original in measuring conventional proximity, it nonetheless offers a much stronger basis for other measures of distance based on accumulation with motion.

The *FocalFUNCTION spreading in* Operation

By introducing the idea that proximity is a quantity that accumulates as a consequence of motion and that it can be measured by simulating that motion, we have shown how several useful variations on *FocalProximity* can be developed. In each of these cases, however, we have dealt with proximity only as a measure of physical separation. At

Figure 5-28 Planar distances between locations defined as nodes in a network. The locations of a layer can be equated with the nodes of a network (white dots) such that each is connected to its adjacent neighbors by lateral links (solid lines) of weight 1.0 and diagonal links (dashed lines) of weight 1.4142. In these terms, the distance between any two locations can then be defined as the total weight of the lowest-weight path of links between them. Here, for example, each location is numbered according to its distance to the location marked 0. Note that the resulting zones of distance form concentric octagons (shaded) rather than circles.

this point, it is useful to adopt a much broader view of distance, one that regards it as a measure of any quantity that can accumulate as a consequence of motion.

Certainly, physical separation is one such quantity. As a snail makes its way across a garden, as an echo makes its way across a canyon, or as an airplane makes its way across a continent, physical distance accumulates, be it in inches, meters, or miles. But so do minutes. And so can dollars, or gallons of fuel, or even levels of environmental impact. For many applications, these units of "distance" will in fact yield measures of relative position that are much more meaningful than conventional measures of physical separation.

To generate zones of distance in units such as time or energy, we can still rely on the wave analogy presented in Figs. 5-25 through 5-27. Note that each wave front in those figures is shown at a position achieved by traveling from the wave source over a series of shortest-possible paths. "Shortest" in this context, however, does not refer to physical distance as measured along a straight line, around a barrier, or over a physical surface. It refers instead to a minimal accumulation of what amount to incremental travel costs associated with different media.

In the case of the waves in the pool of water, these costs relate to time and to energy. As each wave passes from the deeper to the shallower portion of the pool, it expends both time and energy at a relatively higher rate per unit of physical distance traveled. At any given instant, however, each wave will still have reached its current position by accumulating these travel costs at as low a rate as possible.

One way to characterize the nature of a wave medium in terms of its refractive effect is to indicate the number of consecutive wave fronts in that medium per unit of physical distance. Note, for example, in Fig. 5-27 how the number of wave fronts per inch varies from one medium to the other. In this case, each wavelength could well be equated with a constant unit of time (the dripping frequency), and each medium could thereby be characterized in terms of a measure such as seconds per inch or minutes per mile. This measure would be similar to a highway speed limit (or, more precisely, the inverse of a speed limit) in the form of a "friction factor" or an "index of impedance." In more general terms, it would be a measure of how many incremental units of travel cost (in this case, units of time) will be accumulated over a given physical distance.

To make use of this wave analogy in attempting to measure non-Euclidean distance, the concept of motion-transmitting media must be applied to cartographic space. A layer must be created on which different zones are equated with different media, each characterized in terms of the degree to which it will impede some type of motion. Just as in the case of wave media, this can be expressed in terms of the number of incremental units of travel cost that are associated with a standard physical distance. For our purposes, it is convenient to define this standard physical distance as the lateral dimension of a location. By doing so, we can then describe the "width" of each location in terms of minutes of walking time, pennies worth of fuel consumption, number of people encountered, and so on.

This view of cartographic space holds that locations of identical physical dimensions can differ in dimension when measured in terms of a nonEuclidean metric. You may recall that the physical dimension or width of each location on a layer is defined as the layer's resolution. The nonEuclidean width of a location can also be expressed in these terms. In this case, however, it is a form of resolution that varies from one location to another.

To measure distance in these terms, we need only modify the wave-propagating process of operation *FocalFUNCTION spreading* such that links between the nodes of a network that are used to represent locations are weighted not according to geometric length but by incremental travel costs. As illustrated in Fig. 5-29, each link is now weighted by the product of its length and the average of the incremental travel costs of the two locations it connects.

The operation associated with this form of distance measurement is given as

NEWLAYER   = *FocalFUNCTION of FIRSTLAYER at DISTANCE*
                             spreading in FRICTIONLAYER

where *FRICTIONLAYER* is the title of a layer on which each location's value indicates its travel-cost resolution.

An example of this operation is presented in Fig. 5-30. The *ThisTime* layer shown here indicates the distance from each location in the Brown's Pond study area to Brown's Pond itself. Distance is measured not in meters, however, as it is on the *ThisFar* layer shown in Fig. I-2. It is measured instead in minutes of walking time. This measurement assumes a walking pace as indicated on the *Mobility* layer shown in Fig. 4-2: 12 seconds per 20 meters (or 10 minutes per kilometer) on major or minor roads and 36 seconds per 20 meters (or 30 minutes per kilometer) elsewhere. *ThisTime* was generated from *Mobility* and from the *WhichPond* layer shown in Fig. I-1 as follows:

Brown's   = *LocalRating of WhichPond with -0 for 0 2 with 0 for 1*
ThisTime   = *FocalProximity of Brown's spreading in Mobility at ...*

Figure 5-29 Travel-cost distances between locations defined as weighted nodes in a network. The locations of a layer can be equated with the nodes of a network (white dots) such that each is connected to its adjacent neighbors by links (black lines) whose weights reflect increments of cost that accumulate with travel. In these terms, the distance between any two locations can then be defined as the total weight of the lowest-weight path of links between them. Here, for example, each location is numbered according to its travel-cost distance to the location marked 0 given that

- locations in the upper six rows have been set to incremental travel costs of one,
- those below have been set to three, and
- each link's weight is computed by averaging the costs of the two locations it connects and multiplying by 1.4142 if the link is diagonal.

Note how zones of travel-cost distance (shaded) vary in width as a result.

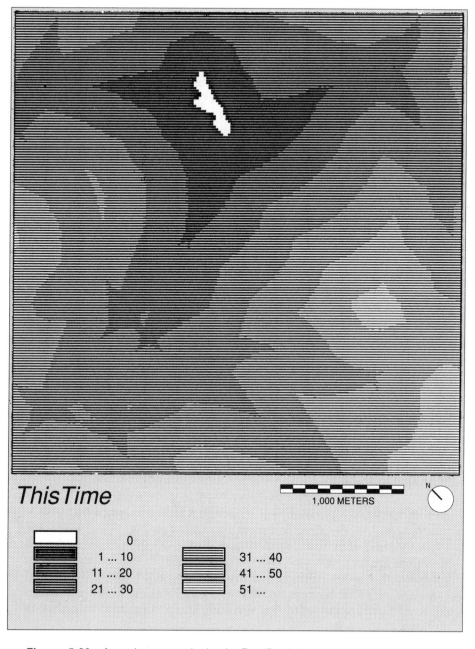

Figure 5-30 A map layer created using the *FocalProximity spreading in* operation. *This-Time* is a layer indicating the estimated walking time in minutes from each location within the Brown's Pond study area to Brown's Pond itself. Note that each shading pattern represents not just one zone but a range of travel times.

The *FocalFUNCTION spreading on* **Operation**

Another useful variation on the *FocalFUNCTION spreading* operation is one that measures distance as illustrated in Fig. 5-31. Note here that incremental distances between adjacent locations are now defined in three rather than two dimensions.

This form of the *FocalFUNCTION* operation is specified as

NEWLAYER = *FocalFUNCTION of FIRSTLAYER at DISTANCE spreading on SURFACELAYER*

where *SURFACELAYER* is the title of a layer on which each location's value indicates its vertical position.

The *FocalFUNCTION spreading through* **Operation**

A further variation on the *FocalFUNCTION spreading* operation is illustrated in Fig. 5-32. Here, the incremental measurements of distance from any given location are restricted to specified directions.

This variation is specified as

NEWLAYER = *FocalFUNCTION of FIRSTLAYER at DISTANCE spreading through NETWORK*

where *NETWORK* is a layer on which each location's value indicates the direction(s) in which increments of distance from that location can be measured. *NETWORK* values represent directions in the same manner as those generated by the *IncrementalDrainage* operation. These are illustrated in Fig. 5-12.

Often, the *NETWORK* layer used in a *FocalFUNCTION spreading through NETWORK* operation will in fact be a layer generated by *IncrementalDrainage*. An example of this is illustrated in Fig. 5-33. Here, the *Upstream* layer shown in Fig. 5-10 and an *EachDrop* layer on which all locations are set to a value of one have been used to estimate the amount of water that would ultimately flow through each location if an equal amount of rainfall were to fall at each location and then drain downhill over the topographic surface. The procedure involved is as follows:

Runoff = *FocalSum of EachDrop spreading through Upstream at ...*
Runoff = *LocalRating of Runoff with 1 for ... 10 with 2 for 11 ... 20 with 3 for 21 ...*

Figure 5-31 Surficial distances between locations defined as nodes in a network. The locations of a layer can be equated with the nodes of a network (white dots) such that each is connected to its adjacent neighbors by links (black lines) whose weights correspond to distances between the centers of grid squares at specified vertical positions. In these terms, the distance between any two locations can then be defined as the total weight of the lowest-weight path of links between them. Here, for example, each location is numbered according to its distance to the location marked 0 given that locations in the lower six rows are at a vertical position 4.899 units higher than those in the upper six rows. As a result, each location in the seventh row is effectively five units (rather than one, since five is the secant of 4.899) from its upper neighbor. Note how this affects zones of distance (shaded).

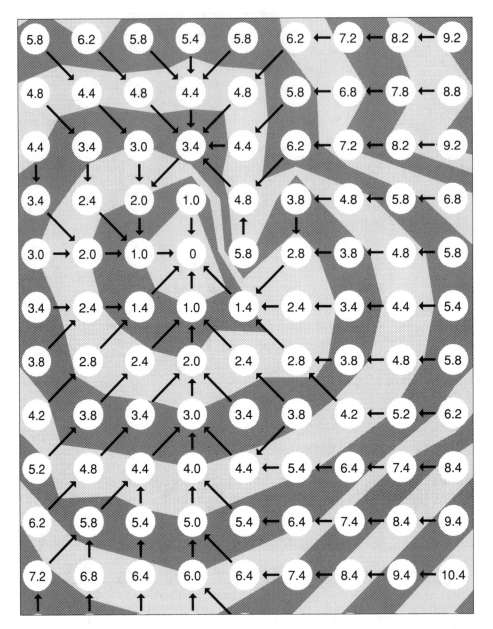

Figure 5-32 Directed distances between locations defined as nodes in a network. The locations of a layer can be equated with the nodes of a network (white dots) such that each is connected by weighted links only to specified neighbors (arrows). In these terms, the distance between any two locations can then be defined as the total weight of the lowest-weight path of links between them. Here, for example, each location is labeled according to its distance to the location marked 0. Note how this affects zones of distance (shaded).

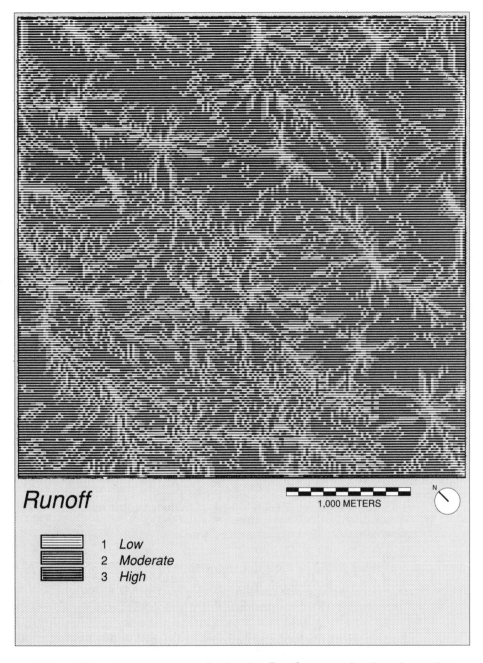

Figure 5-33   A map layer created using the *FocalSum spreading through* operation. *Runoff* is a layer on which each location within the Brown's Pond study area is set to a new value indicating the relative amount of water likely to flow over that location after a rainstorm.

Chapter 5   CHARACTERIZING LOCATIONS WITHIN NEIGHBORHOODS   149

Additional Functions of Extended Neighborhoods

A number of additional functions of extended neighborhoods can also be developed by using measures of distance and/or direction from each neighborhood's focus to define its boundaries and/or weight its values. In general, these functions represent sophisticated methods of cartographic interpolation. They are often associated with the modeling of wave-like phenomena involving noise, light, gravitation, migration, or other forms of spatial diffusion.

These functions can become highly specialized and extremely complex, particularly when factors such as momentum and reflection are introduced. Nonetheless, most will still involve processing similar to that which has been described.

5-3 QUESTIONS

This chapter has presented cartographic modeling capabilities that characterize locations in terms of their surroundings. The following are questions that review and expand on the use of these capabilities.

5-1 How does the adjective *incremental*, as used here, differ from *local* or *focal*?

5-2 How could functions of immediate neighborhoods be used to check the veracity of a recently encoded topographic elevation layer?

5-3 How would you identify the peak locations on that layer of topographic elevations? How about the ridges or the saddle points?

5-4 What would be the effect of operation *IncrementalAspect* on a surface of barometric pressure values? How about *IncrementalDrainage* or *IncrementalGradient*? What would be the effect of these operations on population density surface?

5-5 In New England's glacial landscape, soil depth to bedrock is often shallower on hilltops than in valley bottoms. Assuming that soil depths generally range from one to 10 meters, how could you infer a layer of soil depth from the Brown's Pond *Altitude* layer shown in Fig. 1-9?

5-6 The form of a surface can be accentuated by polarizing elevations such that hills become higher and valleys deeper. How?

5-7 How could a layer of (not upstream but) downstream directions be derived from *IncrementalDrainage* results?

5-8 What is the slope at that point where an incline of zero degrees meets an incline of 45 degrees? Now suppose each of these inclines were to be expressed not as an angle but as a ratio of vertical to horizontal distance. In this more conventional form, the two inclines would measure zero percent and 100 percent, respectively. Does this affect your initial answer and, if so, why? How does it relate to the "best-fitting" planes inferred by operations *IncrementalGradient* and *IncrementalAspect*?

5-9 In 1938, much of Petersham's forest was blown down by a hurricane whose winds had their greatest effect at the northwest edges of forest openings. How could you generate a map layer depicting these wind-sensitive areas?

5-10 How would you generate a layer of road intersections from the *Development* layer shown in Fig. 1-12? How about intersections between roads and streams as depicted on the *Water* layer shown in Fig. 1-10? What about a layer of stream headwaters from *Water* or, from the *Altitude* layer shown in Fig. 1-9, a layer of the headwaters not only for streams but for all surficial flows?

5-11 The effect of operation *FocalMean* on the appearance of a map layer is much like that of blurred vision. Why?

5-12 How do the results of

NEWLAYER   = *FocalMean of NEWLAYER at ... 100*

and

NEWLAYER   = *FocalMean of FIRSTLAYER at ... 50*
NEWLAYER   = *FocalMean of NEWLAYER at ... 50*

5-13 Suppose the *Vegetation* layer shown in Fig. 1-11 were to be supplemented with a similar layer depicting vegetation twenty years earlier. How would you compare these layers to indicate vegetation change in a way that is not affected by the kind of location-specific differences that can be attributed to cartographic imprecision?

5-14 It could be argued that the *HisScore* layer presented in Fig. 4-12 is more "salt-and-peppery" in appearance than the *HerScore* layer shown in Fig. 4-13. How can this argument be substantiated?

5-15 Each of two map layers has a "black" zone and a "white" zone. On one of these layers, transitions from black to white appear to be more gradual than on the other. How can this be quantified?

5-16 The sun is due east at an angle of 10 degrees above the horizon. How would you generate a layer of the shadows it casts on an *Altitude* surface?

5-17 Why would the target locations specified in a *FocalProximity* operation ever be set to anything other than zero?

5-18 Hydrologic pollutants tend to flow downhill. Airborne pollutants tend to flow downwind. Given a layer of areas particularly sensitive to such pollution, how would you generate a layer indicating where not to site a pollution source?

5-19 The best site for an ambulance facility is that which can access the most people in the least time. How would you use *Development* layer shown in Fig. 1-12 to site such a facility in the Brown's Pond study area?

5-20 Deer habitat is best where plentiful food, water, and cover are all within easy access. How would you generate a layer of deer habitat quality that applies these criteria to the Brown's Pond study area?

5-21 The results of a *FocalProximity spreading* operation are subtracted from those of a *FocalProximity spreading in FRIC-TIONLAYER* operation on the same target areas. What does their difference indicate?

5-22 Traffic through the Petersham town center is sometimes restricted during public events and must be routed past Brown's Pond. How could you determine just how much of a detour this entails for those traveling from north to south over major roads?

5-23 How can a layer of watersheds (areas draining to a common location) be derived from a layer of topographic elevations?

5-24 Given the *Altitude* layer shown in Fig. 1-9 and the *Water* layer shown in Fig. 1-10, how would you generate a new layer indicating the number of different ponds potentially visible from each location?

5-25 How could the Brown's Pond *Altitude*, *Vegetation*, and *Development* layers be used to indicate all locations where a new 30-meter building, if constructed, would be visible from one or more existing buildings?

5-26 How could the *FocalInsularity* operation be replicated using *FocalFUNCTION spreading*?

Chapter 6

CHARACTERIZING LOCATIONS
WITHIN ZONES

The third and final group of data-interpreting operations includes those that compute a new value for each location as a function of existing values associated with a zone containing that location. These operations provide for the aggregation of data over units of cartographic space (zones) that are like individual locations to the extent that they provide all-inclusive but mutually exclusive study area coverage. Zones are also like neighborhoods to the extent that they represent two-dimensional areas. Unlike locations, however, zones may vary in their shape and size. And unlike both locations and neighborhoods, their constituent values do not conform to any particular ordering or spatial configuration.

Operations characterizing locations within zones can be classified into two major groups. Respectively, these include
- those operations that characterize entire zones, and
- those that characterize partial zones.

6-1 FUNCTIONS OF ENTIRE ZONES

The first of the two groups of zone-characterizing operations includes those that do so as illustrated in Fig. 6-1. For each location, they compute a new value that summarizes a set of existing values.

These existing values are drawn from all locations that are within the same zone as the initial location on a specified map layer. The existing values can be summarized in terms of zonal maxima, minima, sums, and so on much as values were summarized on a location-by-location basis by *LocalFUNCTION* operations and on a neighborhood-by-neighborhood basis by *FocalFUNCTION* operations. Here, analogous capabilities are expressed in the form of *ZonalFUNCTION* operations.

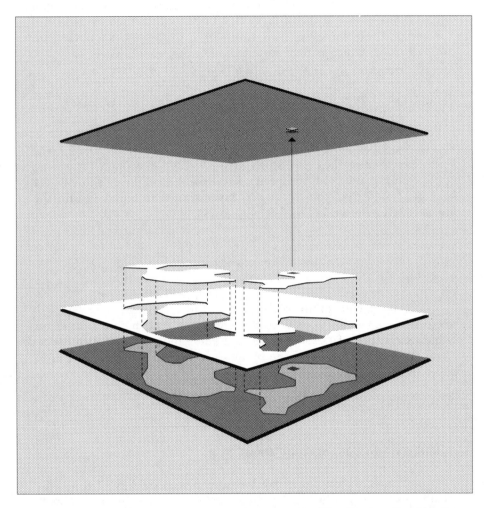

Figure 6-1 Functions of entire zones. Operations *ZonalCombination*, *ZonalMajority*, *ZonalMaximum*, *ZonalMean*, *ZonalMinimum*, *ZonalMinority*, *ZonalProduct*, *ZonalRating*, *ZonalSum*, and *ZonalVariety* can all be used to compute a new value (above) for each location as a specified function of existing values within a common zone (below).

The *ZonalCombination, ZonalMajority , ZonalMaximum, ZonalMean , ZonalMinimum, ZonalMinority, ZonalProduct, ZonalRating , ZonalSum*, **and** *ZonalVariety* **Operations**

The values of locations within zones can be summarized by way of functions corresponding to any of those forms of *LocalFUNCTION* or *FocalFUNCTION* that are commutative. These functions are embodied in operations given as *ZonalCombination, ZonalMajority, ZonalMaximum, Zonal-Mean, ZonalMinimum, ZonalMinority, ZonalProduct, ZonalRatng, ZonalSum*, and *Zonal-Variety*. Each generates a new layer by first summarizing the values of all locations on one existing layer that occur within each of the zones on a second existing layer. This summary is expressed as a new value for each zone, which is then assigned all of that zone's locations.

The *ZonalFUNCTION* operation is specified as

NEWLAYER   = ZonalFUNCTION of FIRSTLAYER [within SECONDLAYER]

where *FIRSTLAYER* is the layer of values to be summarized and *SECOND-LAYER* the layer whose zones aggregate those values. If no *SECOND-LAYER* is specified, the entire study area is treated as a single zone.

In Fig. 6-2 is an example. The *HomesPerBlock* layer shown here was generated from the *Development* layer shown in Fig. 1-12 and the *EachBlock* layer shown in Fig. 5-5 as follows:

Housing   = LocalRating of Development with 0 for ... with 1 for 3
HomesPerBlock   = ZonalSum of Housing within EachBlock
HomesPerBlock   = LocalRating of HomesPerBlock with 0 for -0 with 10 for 10 ...

Another example of *ZonalFUNCTION* is presented in Fig. 6-3. Here, a *ZonalMinimum* operation has been used to generate a layer on which each zone of *EachBlock* is set to a value indicating its distance to Brown's Pond. This was done by aggregating values from the *ThisFar* layer shown in Fig. I-2 as follows:

BlockProximity   = ZonalMinimum of ThisFar within EachBlock

Additional Functions of Entire Zones

Values within zones can be summarized by way of functions analogous to any of the commutative functions of locations or neighborhoods. To replicate these functions generally involves the use of a *Local-FUNCTION* operation to combine the results of two or more *ZonalFUNCTION* operations.

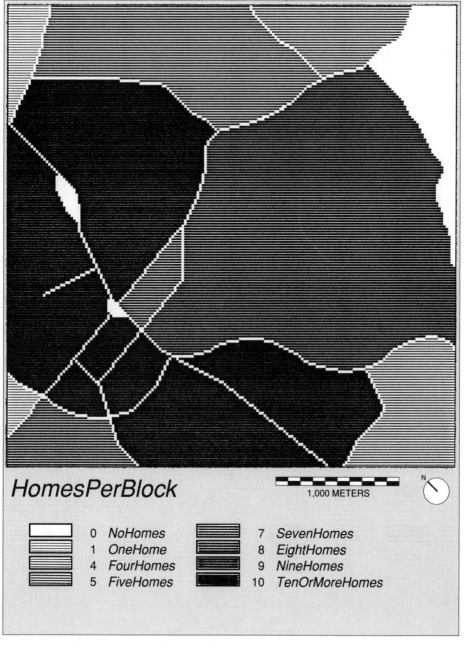

Figure 6-2 A map layer created using the *ZonalSum* operation. *HomesPerBlock* is a layer indicating the number of homes within each conterminous block of land between roads in the Brown's Pond study area.

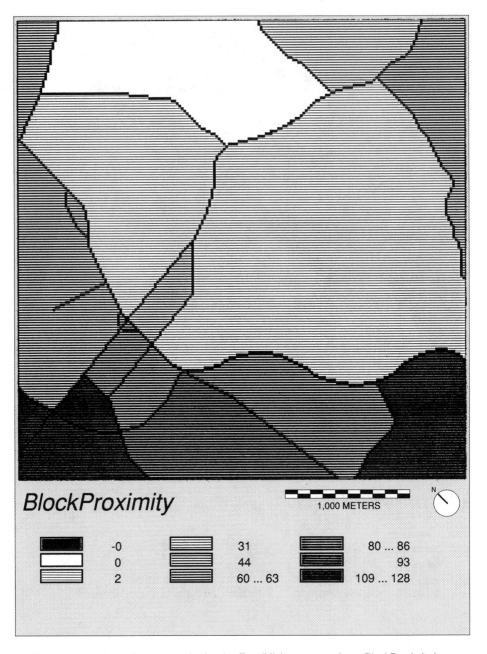

Figure 6-3 A map layer created using the *ZonalMinimum* operation. *BlockProximity* is a layer on which each location within the Brown's Pond study area is set to a value indicating the shortest distance to Brown's Pond from any location in the same block of non-road land. Note that some shading patterns represent not just one zone but a range of distances.

6-2 FUNCTIONS OF PARTIAL ZONES

The second of the two groups of operations characterizing locations within zones includes those that contrast an existing value of each location to a statistic summarizing the values of all locations in a similar zone. As illustrated in Fig. 6-4, this is generally done by
- comparing two layers to determine which locations of one lie within which zones of the other,
- summarizing the values of those locations within each of those zones,
- contrasting the initial value of each location to the summary statistic for its zone, and
- recording that contrast as the location's new value.

These operations include *ZonalPercentage*, *ZonalPercentile*, and *ZonalRanking*.

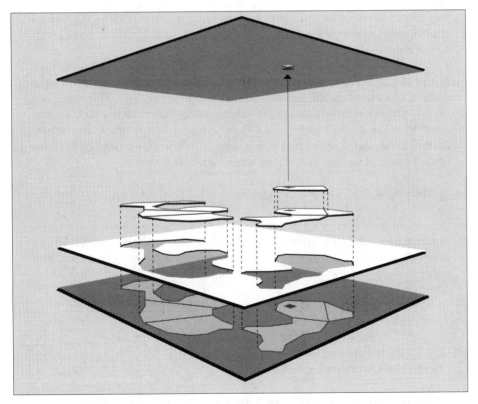

Figure 6-4 Functions of partial zones. Operations *ZonalPercentage*, *ZonalPercentile*, and *ZonalRanking* can all be used to compute a new value for each location (above) as a function contrasting its existing value (center) to those of others within its zone (below).

The *ZonalPercentage*, *ZonalPercentile*, **and** *ZonalRanking* **Operations**

The *ZonalPercentage*, *ZonalPercentile*, and *ZonalRanking* operations are specified much like other *ZonalFUNCTION* operations. Their statements are given as

NEWLAYER   = ZonalFUNCTION of FIRSTLAYER [within SECONDLAYER]

These operations contrast the values of individual locations to the values that occur within their zones in much the same way as operations *FocalPercentage*, *FocalPercentile*, and *FocalRanking* do so within neighborhoods. *ZonalPercentage* assigns a new value to each location within a *SECONDLAYER* zone indicating what percent of that zone shares the location's *FIRSTLAYER* value. *ZonalPercentile* indicates the percentage of each location's *SECONDLAYER* zone that is occupied by locations of lower *FIRSTLAYER* value. And *ZonalRanking* characterizes each location in terms of the number of zones having lower *FIRSTLAYER* values that occur (either fully or partially) within that location's *SECOND-LAYER* zone.

An example of *ZonalRanking* is presented in Fig. 6-5. Here, values of the *ThisFar* layer shown in Fig. I-2 have, in effect, been sorted within each of the zones of the *EachBlock* layer shown in Fig. 5-5. The result is a layer entitled *ProximityByBlock* on which each individual block's locations that are closest to Brown's Pond (regardless of what that actual distance may be) are set to a value of one, while those next closest are set to two, and so on. This layer was created as follows:

ProximityByBlock = ZonalRanking of ThisFar within EachBlock

In Fig. 6-6 is an example of operation *ZonalPercentage*. Here, *EachBlock* has been combined with the *HisScore* layer shown in Fig. 4-12 to generate a layer indicating where and how much they coincide. The value assigned to each location on *HisScoreVsBlock* is computed as an average of
- the percentage of that location's *EachBlock* zone that shares its *HisScore* value, and
- the percentage of the location's *HisScore* zone that shares its *EachBlock* value.

These values were generated as follows:

OverlapPerBlock   = ZonalPercentage of HisScore within EachBlock
OverlapPerScore   = ZonalPercentage of EachBlock within HisScore
HisScoreVsBlock   = LocalMean of OverlapPerBlock and OverlapPerScore

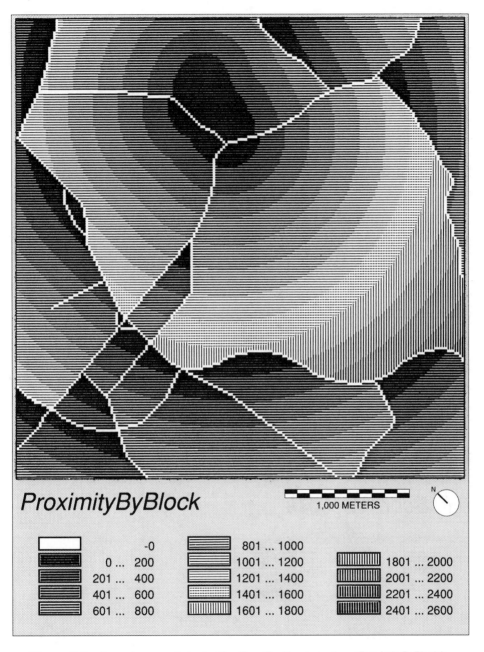

ProximityByBlock

1,000 METERS

N

	-0
	0 ... 200
	201 ... 400
	401 ... 600
	601 ... 800

	801 ... 1000
	1001 ... 1200
	1201 ... 1400
	1401 ... 1600
	1601 ... 1800

	1801 ... 2000
	2001 ... 2200
	2201 ... 2400
	2401 ... 2600

Figure 6-5 A map layer created using the *ZonalRanking* operation. *ProximityByBlock* is a layer indicating, within each conterminous block of non-road land, the relative magnitude of values representing proximity to Brown's Pond in meters. Note that each shading pattern represents not just one zone but a range of distances.

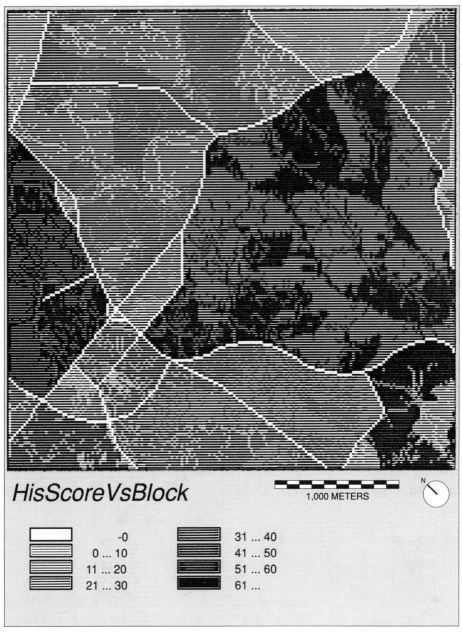

HisScoreVsBlock

1,000 METERS

N

-0	31 ... 40
0 ... 10	41 ... 50
11 ... 20	51 ... 60
21 ... 30	61 ...

Figure 6-6 A map layer created using the *ZonalPercentage* operation. *HisScore-VsBlock* is a layer on which each location within the Brown's Pond study area is set to a value indicating the amount of coincidence between its zones on two different layers. Note that each shading pattern represents not just one zone but a range of coincidence values.

Additional Functions of Partial Zones

A number of additional functions can also be developed by contrasting the values of individual locations within zones to zonal statistics. Many of these functions can be replicated by using a *ZonalFUNCTION of FIRSTLAYER* operation followed by a *LocalFUNCTION* to combine the *ZonalFUNCTION* result with the *FIRSTLAYER* from which that result was created.

Consider, for example, the following procedure:

Highest	= *ZonalMaximum of FIRSTLAYER*
Lowest	= *ZonalMinimum of FIRSTLAYER*
Range	= *LocalDifference of Highest and Lowest*
Interval	= *LocalRatio of Range and 100*
HowHigh	= *LocalDifference of FIRSTLAYER and Lowest*
NEWLAYER	= *LocalRatio of HowHigh and Interval*

Here, locations are aggregated into groups defined by equally spaced intervals of *FIRSTLAYER* value. Those groups are then represented by *NEWLAYER* values ranging from zero to 100. While the result is much like that which could have been generated with *LocalRating*, the procedure is nonetheless distinctly zonal rather than local in nature.

6-3 QUESTIONS

This chapter has introduced cartographic modeling capabilities that characterize locations within zones. Below are questions that relate to the use of these zone-characterizing capabilities.

6-1 The normalization procedure presented above is described as "distinctly zonal rather than local in nature." Why?

6-1 If every homeowner in the Brown's Pond study area were to commute to the north, that highway detour described in Ques. 5-22 would increase average commuting time. How could this be measured?

6-2 How could the Brown's Pond *Altitude* layer be revised to indicate elevations at the surface rather than at the bottom of ponds?

6-3 The *LocalCombination* operation is, to some extent, also zonal in nature. How? What about *FocalInsularity*?

6-4 Given the *Water* layer shown in Fig 1-10, how could you effectively erase all ponds of less than ten hectares?

6-5 Hills in Petersham tend to be oriented along a north-south axis. How can this observation be supported by comparing layers of topographic slope and aspect? How could a similar procedure be used to support an observation that housing density tends to increase with proximity to the Petersham town commons, or that most of the homes in Petersham are along the minor roads?

6-6 The *ThisScore* layer shown in Fig. 4-10 clearly reflects its origins in the Brown's Pond *Altitude*, *Water*, *Vegetation*, and *Development* layers. How can its relationship to each of these layers be more precisely examined? How could a similar technique be used to explain the existing pattern of land development in Petersham?

6-7 Which of the zones depicted on the *EachBlock* layer in Fig. 5-5 is most "central" to the Brown's Pond study area? How can you demonstrate this?

6-8 Given a layer of forest heights and a layer of land ownership, how would you locate each property's tallest tree(s)? How about the highest elevation(s) above sea level in the Brown's Pond study area?

6-9 A golf course is to be sited in the Brown's Pond study area on a parcel of at least 100 hectares that is unbroken by streams or roads and has as much topographic variety as possible. How should such a site be selected?

6-10 Suppose that those deer mentioned in Ques. 5-20 are willing and able to travel anywhere as long as they need not cross a road. How does this affect that assessment of deer habitat quality?

6-11 The Petersham Center School is within zone 13 on the *EachBlock* layer shown in Fig. 5-5. How would you calculate, for each home in the Brown's Pond study area, the minimum number of roads that must be crossed in walking to school?

6-12 How would you compute the variance of one layer's values within each of another layer's zones?

6-13 What would be the local and focal analogues to what are here referred to as *partial zones*?

Part III

CARTOGRAPHIC MODELING TECHNIQUES

Chapter 7

DESCRIPTIVE MODELING

Given a set of basic cartographic modeling capabilities, we can now develop a variety of more sophisticated techniques. This is generally a matter of combining selected operations into procedures that are tailored to the needs of particular applications. At the beginning of this process, modeling efforts tend to be *descriptive*. They attempt to describe in geographic terms "what is" or perhaps "what could be." Later, modeling often tends to shift to concern for "what should be," moving from the descriptive to a more *prescriptive* intent.

This chapter and the one that follows respectively introduce techniques for
- descriptive and
- prescriptive

cartographic modeling. Though they certainly do not exhaust all possibilities, they do begin to suggest a common approach to a wide range of problems.

Among those cartographic modeling techniques whose purpose is to describe, a broad distinction can be drawn between those that analyze and those that synthesize cartographic data. *Analytic* techniques decompose data into finer levels of meaning, while *synthetic* techniques recompose data for use in particular contexts.

Techniques for the analysis and synthesis of cartographic data have much in common with general statistical methods. What distinguishes the cartographic techniques is their ability to deal with observations that are spatially interrelated. For our immediate purposes, we need not venture far into the substantial field of spatial statistics. However, we must develop basic techniques for cartographic description. Among these are techniques associated with
- the analysis of cartographic position,
- the analysis of cartographic form, and
- the synthesis of cartographic characteristics.

7-1 ANALYSIS OF CARTOGRAPHIC POSITION

As indicated earlier, the position of a cartographic condition is usually expressed through measurements of distance and/or direction. These measurements may describe the condition in either of two ways. They may indicate its *absolute* position with respect to the column-row frame of reference that is shared by all locations, or they may indicate its *relative* position with respect to a particular set of locations.

Characterizing Absolute Position

To characterize the position of a cartographic condition with respect to its absolute frame of reference requires only that each of the locations involved be set to the values of its two coordinates. This can generally be done by using the data preparation capabilities of a geographic information system to generate one map layer on which each location is set to the value of its column coordinate and another on which each location is set to the value of its row coordinate. As suggested in Ques. 4-7, absolute position can then be treated like any other local characteristic.

To illustrate, consider the task of locating zonal *centroids*. A centroid is that one location around which a zone's area is evenly "balanced" in all directions. Below is a procedure that uses an *EachColumn* layer of column coordinates and an *EachRow* layer of row coordinates to generate a *NEWLAYER* indicating the centroids of *FIRSTLAYER* zones.

CenterColumn	=	*ZonalMean of EachColumn within FIRSTLAYER*
CenterRow	=	*ZonalMean of EachRow within FIRSTLAYER*
ColumnShift	=	*LocalDifference of CenterColumn and EachColumn*
RowShift	=	*LocalDifference of CenterRow and EachRow*
NEWLAYER	=	*LocalRating of ColumnShift and RowShift*
		with FIRSTLAYER for 0 0

Column and row coordinates can also be used to shift values from one location to another in ways that scale (enlarge or reduce), translate (move parallel to columns and/or rows), rotate, or otherwise reproject a cartographic image with respect to its locational grid. The following, for example, is a procedure that uses the *EachColumn* and *EachRow* layers described above to rotate an existing *FIRSTLAYER* about a location whose column and row coordinates are respectively given as *PIVOTCOLUMN* and *PIVOTROW*. The rotation is counterclockwise at an angle whose sine and cosine are given as *SINE* and *COSINE*, respectively.

HowFarRight	=	LocalDifference of EachColumn and PIVOTCOLUMN
HowFarUp	=	LocalDifference of EachRow and PIVOTROW
NewColumn	=	LocalProduct of HowFarRight and COSINE
NewRow	=	LocalProduct of HowFarUp and SINE
ColumnShift	=	LocalDifference of NewColumn and NewRow and HowFarRight
NewColumn	=	LocalProduct of HowFarRight and SINE
NewRow	=	LocalProduct of HowFarUp and COSINE
RowShift	=	LocalDifference of NewRow and NewColumn and HowFarUp
Bearing	=	LocalArcTangent of RowShift and ColumnShift
RowShift	=	LocalProduct of RowShift and RowShift
ColumnShift	=	LocalProduct of ColumnShift and ColumnShift
Range	=	LocalSum of RowShift and ColumnShift
Range	=	LocalRoot of Range and 2
NEWLAYER	=	FocalNeighbor of FIRSTLAYER at Range by Bearing

Characterizing Relative Position

To characterize cartographic position relative to an arbitrary set of locations ultimately requires that all positions involved be related to a common frame of reference. Relative positions are typically expressed, however, not in terms of absolute coordinates but in terms of the distances and/or directions between locations.

As indicated in Sec. 5-2, the distance from a cartographic condition to surrounding locations can generally be regarded as a measure of some quantity that has accumulated as a consequence of motion. Indicative of this is an analogy that can be drawn between the *FocalProximity spreading* operation and mathematical integration. Just as the integral of a mathematical function can be regarded as a measure of the area under that function's curve between two of the points on its independent axis, each of the values generated by a *FocalProximity spreading in FRICTIONLAYER* operation can be regarded as a measure of the area under that *FRICTIONLAYER*'s surface along a minimum-cost path.

This analogy can be taken one step further by noting that the inverse of mathematical integration is differentiation and that the first derivative of a mathematical function is a measure of the slope of that function's curve at any given point. The cartographic analogue to this would be *IncrementalGradient*. If truly analogous, *IncrementalGradient* should serve as the inverse of *FocalProximity spreading in FRICTIONLAYER*. The extent to which it does so is shown in Fig. 7-1. Here, *IncrementalGradient* has been applied to the *ThisTime* layer shown in Fig. 5-30. Note how the resulting *InferredMobility* layer reflects the *Mobility* layer, shown in Fig. 4-2, from which *ThisTime* was created.

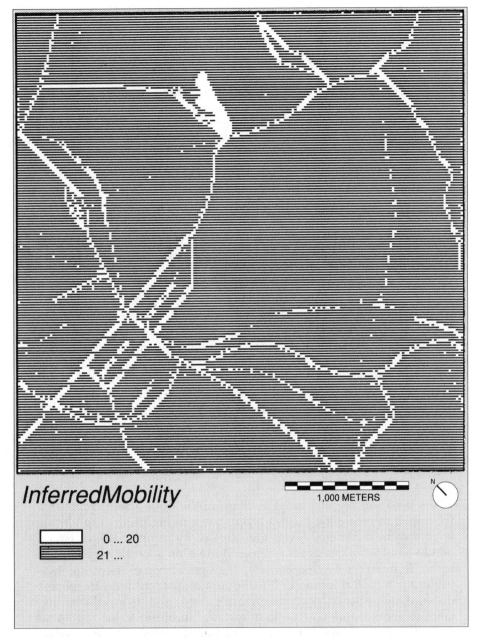

Figure 7-1 A map layer created using the *IncrementalGradient* operation to invert the effect of *FocalProximity spreading in*. *InferredMobility* is a layer indicating the change in cumulative travel time to Brown's Pond over each location within the Brown's Pond study area. Note that each shading pattern represents not just one zone but a range of values.

Application of the *IncrementalGradient* operation to measurements of proximity is also significant in another way. It points to the fact that layers of proximity values are surficial. As such, these layers are also amenable to other operations more typically associated with topographic surfaces.

Consider, for example, the effect of applying *IncrementalDrainage* to a surface of proximity values. One instance of this is illustrated in Fig. 7-2. By comparing the *ThisWay* layer shown here to the *Mobility* and *ThisTime* layers respectively shown in Figs. 4-2 and 5-30, it can be seen that each *ThisWay* value indicates the direction of the path of minimum walking time to Brown's Pond. To understand why this is the case, consider the meaning of "steepest descending slope" (the basis on which *IncrementalDrainage* determines direction) in the present context. From any location on a travel-cost distance surface, the direction of steepest descent will be that which achieves the greatest reduction in travel cost per unit of physical distance traveled.

By applying additional operations to the kind of distance values that are generated by *FocalProximity spreading in*, the path(s) of minimum cost to a specified condition can actually be traced. In Fig. 7-3, for example, is a layer that was generated from *ThisWay* as follows:

BestPath   = *FocalSum of EndPoint spreading through ThisWay at ...*

where *EndPoint* is a layer on which a selected location at the right edge of the study area is set to a value of one, while all others are set to zero.

An equally effective and sometimes more useful technique for identifying minimum-cost paths is shown in Figs. 7-4 and 7-5. The *ThatTime* layer shown in Fig. 7-4 was generated as follows:

ThatTime   = *FocalProximity of EndPoint spreading in Mobility at ...*

where distances are measured in terms of the same *Mobility* units that were used to create the *ThisTime* layer shown in Fig. 5-30. The *BestTime* layer presented in Fig. 7-5 was then created by using a *LocalSum* operation to add the values of *ThisTime* and *ThatTime* on a location-by-location basis. To interpret *BestTime*, consider the significance of the value it attributes to a typical location. It is the sum of that location's travel-cost distance to one set of locations (Brown's Pond) and its travel-cost distance to another (the *EndPoint* location). As such, this value represents the lowest possible total cost of any path that connects those two destinations by passing through the intervening location. This not only serves to delineate the minimum cost path but also gives a useful indication of higher cost alternatives.

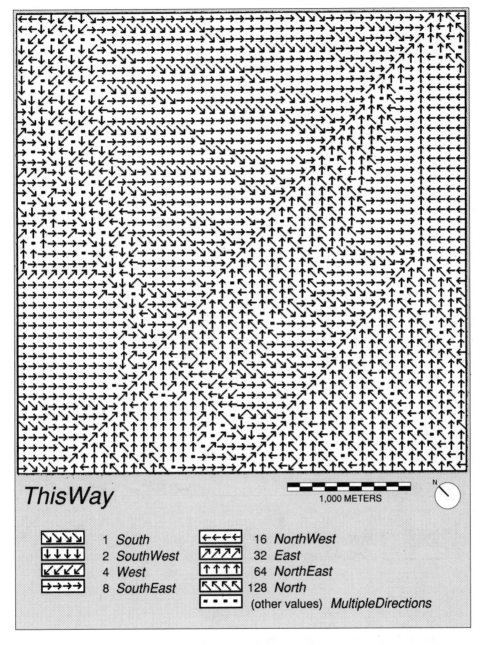

ThisWay

1,000 METERS

N

↘↘↘↘	1 *South*	←←←←	16 *NorthWest*
↓↓↓↓	2 *SouthWest*	↗↗↗↗	32 *East*
↙↙↙↙	4 *West*	↑↑↑↑	64 *NorthEast*
→→→→	8 *SouthEast*	↖↖↖↖	128 *North*
		▪ ▪ ▪ ▪	(other values) *MultipleDirections*

Figure 7-2 A map layer created by applying *IncrementalDrainage* to a layer generated by *FocalProximity spreading in*. *ThisWay* is a layer indicating directions of minimum travel cost toward Brown's Pond. Here, only a portion of the Brown's Pond study area (just left of center in the vicinity of the town commons) is shown.

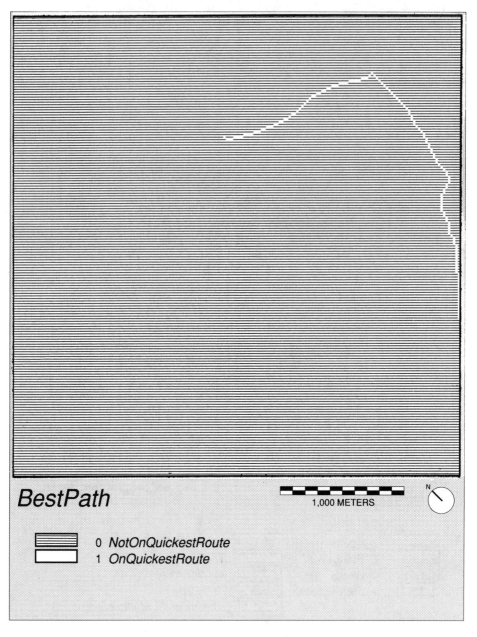

BestPath

1,000 METERS

N

0 *NotOnQuickestRoute*
1 *OnQuickestRoute*

Figure 7-3 A map layer created by applying the *FocalSum spreading through* operation to a layer generated by applying *IncrementalDrainage* to a surface generated with *FocalProximity spreading in*. *BestPath* is a layer indicating the quickest walking route between Brown's Pond and a selected location at the southeast edge of the Brown's Pond study area.

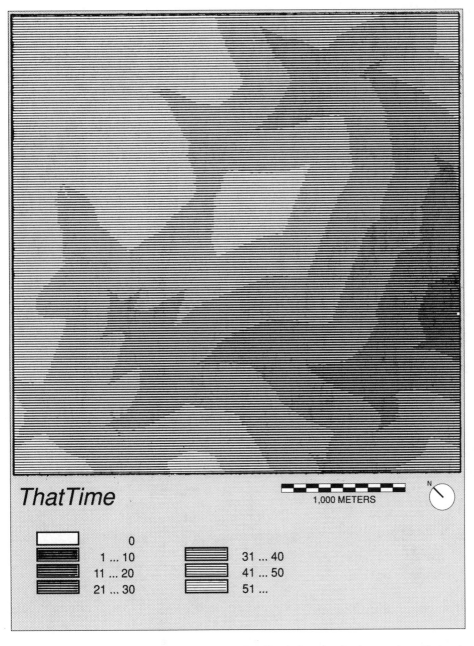

Figure 7-4 A map layer created using the *FocalProximity spreading in* operation. *That-Time* is a layer indicating the estimated walking time in minutes from each location within the Brown's Pond study area to a particular location on its southeast border. Note that each shading pattern represents not just one zone but a range of walking times.

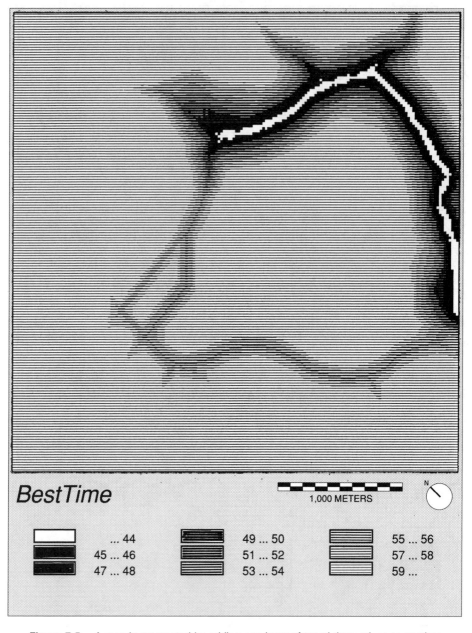

BestTime

1,000 METERS

N

... 44	49 ... 50	55 ... 56
45 ... 46	51 ... 52	57 ... 58
47 ... 48	53 ... 54	59 ...

Figure 7-5 A map layer created by adding one layer of travel time values to another. *BestTime* is a layer on which each location within the Brown's Pond study area is set to a value indicating the minimum time required to walk from Brown's Pond through that location and on to a specified location at the southeast edge of the study area. Note that each shading pattern represents not just one zone but a range of walking times.

7-2 ANALYSIS OF CARTOGRAPHIC FORM

In contrast to the measurements of distance or direction that may be used to characterize the position of a cartographic condition, the form of such a condition is expressed through measurements of shape and size. These measurements reflect distance and directional relationships that are embodied within the condition itself. As indicated in Sec. 1-9, such measurements will vary according to whether the cartographic form is
- punctual,
- lineal,
- areal, or
- surficial.

Characterizing Punctual Form

If a condition is truly punctual in nature, it will have no measurable shape or size and therefore no form to be characterized. A collection of punctual conditions such as a cluster of sampling points or a network of transportation depots, however, may well exhibit non-punctual qualities when considered as a whole. These qualities may be lineal, areal, or even surficial in nature depending on the conditions involved.

Characterizing Lineal Form

If a cartographic condition is lineal in nature, its form can be analyzed in terms of characteristics relating to both shape and size. As indicated earlier, it is convenient to represent a lineal condition as a string of consecutive locations and to infer that these locations are connected to one another as shown in Fig. 1-17. This is the inference drawn by the *IncrementalLinkage* and *IncrementalLength* operations.

Once the form of a lineal condition has been characterized by an operation such as *IncrementalLinkage* or *IncrementalLength*, the incremental measures that result can then be aggregated within each of the zones from which they were generated. In this way, the overall shape or size of a lineal condition can be measured. The following, for example, is a procedure that uses the *Mobility* layer shown in Fig. 4-2 and the *Altitude* layer shown in Fig. 1-9 to generate a layer indicating the length of each stretch of roadway between intersections in the Brown's Pond road network.

RoadForm	= IncrementalLinkage of Mobility
AllRoads	= LocalRating of RoadForm
	with -0 for ... with 1 for 1 ... 28
EachRoad	= FocalInsularity of AllRoads
RoadLength	= IncrementalLength of EachRoad on Altitude
EachLength	= ZonalSum of RoadLength within EachRoad

Characterizing Areal Form

The overall shape and/or size of an areal condition can also be characterized by aggregating incremental measures. One example of this is presented in Fig. 7-6. Here, measurements of frontage and area for individual locations have been aggregated by zone to generate a measure of each zone's *areal roundness*. This is the degree to which the overall shape of a zone curves outward rather than inward. Roundness can be computed as 354 times the square root of a zone's area divided by the length of its perimeter. This figure will range from a maximum of approaching 100 for a circle to a minimum approaching zero as shapes become highly contorted. The procedure used to generate the layer shown in Fig. 7-6 is as follows:

EveryField	= LocalRating of Vegetation with -0 for 1 2 3
EachField	= FocalInsularity of EveryField
Area	= IncrementalArea of EachField
Area	= ZonalSum of Area within EachField
Area	= LocalRoot of Area and 2
Area	= LocalProduct of Area and 354
Perimeter	= IncrementalFrontage of EachField
Perimeter	= ZonalSum of Perimeter within EachField
HowRound	= LocalRatio of Area and Perimeter

To illustrate the utility of this kind of measure, imagine a regional landscape composed of areas that are designated as urban or rural and that occur in equal proportions. Is this a landscape of urban centers on a rural background or one of rural enclaves in an urban context? If it were a landscape of land and water, would it be one of islands or pools? In each case, the answer lies not in proportions but in roundness.

Another way in which the overall form of an areal zone can be characterized by aggregating incremental measures is in terms of its *topological genus*. This is the number of "holes" (conterminous areas of foreign value) that lie within the zone, plus one minus the number of "fragments" (conterminous pieces) that comprise the zone itself. A pair

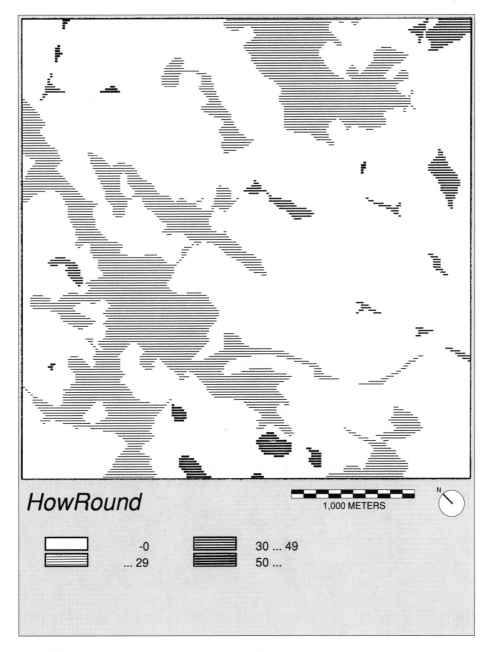

Figure 7-6 A map layer indicating areal roundness. *HowRound* is a map layer on which each forest opening within the Brown's Pond study area is set to a value indicating the degree to which its areal form is round. Note that each shading pattern represents not just one zone but a range of values.

of figure eights, for example, would be of topological genus three.

Genus can be computed as shown in Fig. 7-7. The *HowHoley* layer shown in Fig. 7-8 was generated by applying this technique to the *Each-Field* layer mentioned above as follows:

UpperRightCorner = *IncrementalPartition of EachField*

LowerRightCorner = *FocalRating of UpperRightCorner at 20   by 225*

UpperLeftCorner = *FocalRating of UpperRightCorner at 20   by 335*

LowerLeftCorner = *FocalRating of UpperRightCorner at   20 ... 40 by 270*

UpperRightCorner = *LocalRating of UpperRightCorner with 0 for ...*
    with -1 for 52 7 12 8 9 14 with 1 for 3 13

LowerRightCorner = *LocalRating of LowerRightCorner with 0 for ...*
    with -1 for 4 7 13 11 9 14 with 1 for 1 12

UpperLeftCorner = *LocalRating of UpperLeftCorner with 0 for ...*
    with -1 for 1 7 13 8 10 14 with 1 for 4 12

LowerLeftCorner = *LocalRating of LowerLeftCorner with 0 for ...*
    with -1 for 3 7 12 10 11 14 with 1 for 2 13

UpperRightCorner = *ZonalSum of UpperRightCorner within EachField*

LowerRightCorner = *ZonalSum of LowerRightCorner within EachField*

UpperLeftCorner = *ZonalSum of UpperLeftCorner within EachField*

LowerLeftCorner = *ZonalSum of LowerLeftCorner within EachField*

HowHoley   = *LocalSum of UpperRightCorner and LowerRightCorner*
    and UpperLeftCorner and LowerLeftCorner and 4

HowHoley   = *LocalProduct of HowHoley and 4*

HowHoley   = *LocalRating of HowHoley with 1 for 1 ...*

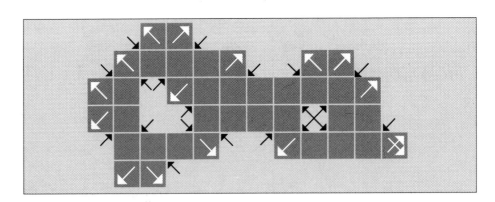

Figure 7-7 Incremental calculation of topological genus. The number of holes within a conterminous figure comprised of grid squares can be calculated incrementally by adding four to the number of inward-pointing right angles around its perimeter(s) (black arrows), subtracting the number of outward-pointing angles (white arrows), and dividing by four. Here, for example, with 20 inward angles and 16 outward angles: (4 + 20 - 16) / 4 = 2.

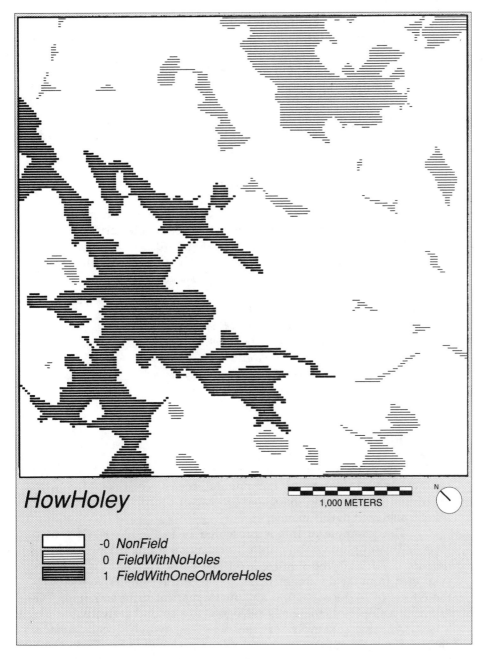

Figure 7-8 A map layer indicating topological genus. *HowHoley* is a map layer on which each forest opening within the Brown's Pond study area is set to a value indicating its topological genus, the number of "holes" or "islands" of forest vegetation within it.

Note that this genus-measuring technique relies on measures of *areal protrusion* and *intrusion* at the scale of individual locations. The technique illustrated in Fig. 7-6 also relies on measures of areal protrusion and intrusion, but at the scale of entire zones. Fig. 7-9 illustrates another technique that deals with the protrusion and intrusion of areal forms but does so at intermediate scales. Here, each location within the open land shown on *EachField* has been characterized according to the proportion of its neighborhood that is also in open land. The resulting *HowPeculiar* layer, computed as

HowPeculiar	=	*FocalPercentage of EachField at ... 100*

tends to accentuate those portions of an areal figure that protrude.

A related technique can be used to measure the narrowness of a zone. The logic involved is illustrated in Fig. 7-10.

An example of this is illustrated in Fig. 7-11. The *HowNarrow* layer shown here was generated from *EachField* as follows:

NonField	=	*LocalRating of EachField with -0 for ... with 1 for -0*
HowFarIn	=	*FocalProximity of NonField at ... 60*
Nucleus	=	*LocalRating of HowFarIn with -0 for ... 60 with 0 for 60 ...*
HowFarOut	=	*FocalProximity of Nucleus at ... 60*
HowNarrow	=	*LocalRating of EachField and HowFarOut with 1 for 1 on ... 60 with 2 for 1 on 60 ...*

Characterizing Surficial Form

Just as the overall form of a lineal or an areal condition can be characterized in terms of incremental measures associated with individual locations, the shape and/or size of a surficial condition can also be analyzed in this manner.

One example of this is presented in Fig. 7-12. Here, a map layer indicating topographic inflection has been created by applying *IncrementalGradient* to the *Steepness* layer shown in Fig. 5-8. As might be suspected from the discussion in Sec. 7-1 regarding similarities between the effects of *IncrementalGradient* and mathematical differentiation, this "slope of a slope" calculation is analogous to a second derivative.

Another example of the way in which local surface-characterizing operations can be combined is shown in Fig. 7-13. Here, topographic slope and aspect values have been combined to generate a layer of shaded relief. *HowShady* was created by reorienting the *Altitude* layer shown in Fig. 1-9 from 45 to -45 degrees (to face a rising sun) and

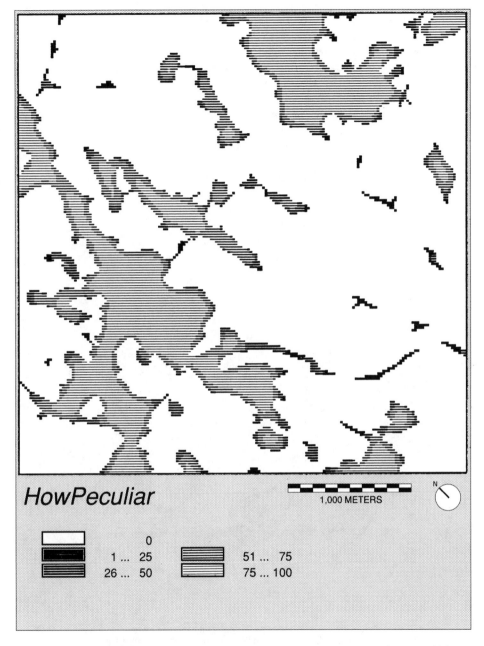

Figure 7-9 A map layer indicating areal protrusion. *HowPeculiar* is a map layer on which each nonforest location within the Brown's Pond study area is set to a value indicating what percent of its neighborhood is also in open land. Note that each shading pattern represents not just one zone but a range of percentages.

then applying a procedure given as follows:

Bearing	=	IncrementalAspect of Altitude
ReverseBearing	=	LocalDifference of 360 and Bearing
Bearing	=	LocalMinimum of Bearing and ReverseBearing
Bearing	=	LocalCosine of Bearing
Slope	=	IncrementalGradient of Altitude
Slope	=	LocalArcTangent of Slope
Slope	=	LocalProduct of Slope and Bearing
SunAngle	=	LocalSum of Slope and 30
SunAngle	=	LocalSine of SunAngle
Slope	=	LocalCosine of Slope
Slope	=	LocalRatio of 20 and Slope
HowShady	=	LocalProduct of SunAngle and Slope
HowShady	=	ZonalPercentile of HowShady
HowShady	=	LocalRatio of HowShady and 9

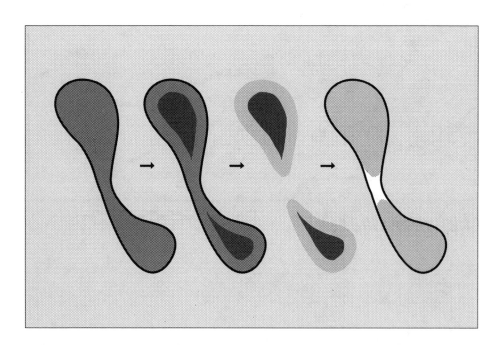

Figure 7-10 Calculation of narrowness. To calculate the narrowness of a zone at each of its locations, all locations beyond a specified distance from any other zone are first identified as the zone's nucleus (dark gray). Each location's distance to that nucleus is then measured. If this distance is less than the distance initially used to define that nucleus (lighter gray), then the location must be part of an area in which the minimum width of the zone is wider than twice the distance. Otherwise (white), it is narrower.

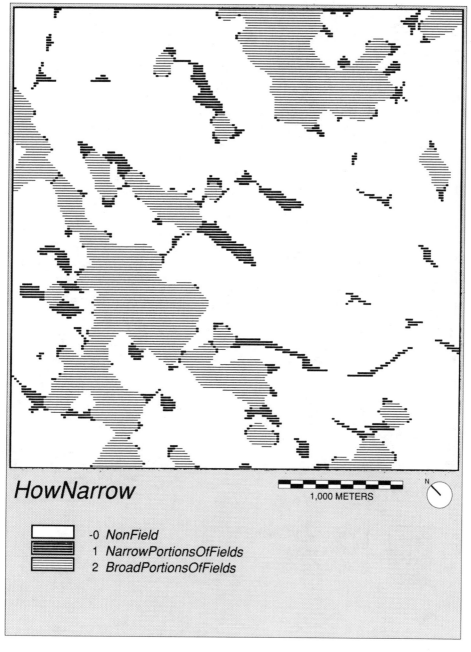

HowNarrow

1,000 METERS

N

- -0 *NonField*
- 1 *NarrowPortionsOfFields*
- 2 *BroadPortionsOfFields*

Figure 7-11 A map layer indicating areal narrowness. *HowNarrow* is a map layer on which each location within the Brown's Pond study area is characterized according to the width of the forest opening, if any, at that location.

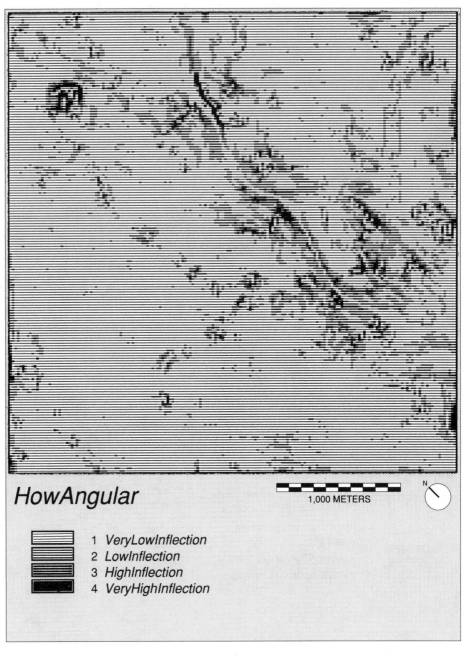

Figure 7-12 A map layer indicating topographic inflection. *HowAngular* is a map layer indicating the degree to which the topographic surface bends at each location within the Brown's Pond study area.

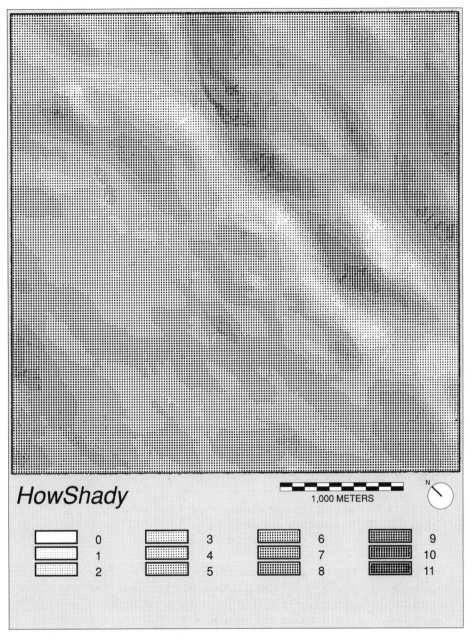

Figure 7-13 A map layer indicating topographic insolation. *HowShady* is a map layer on which each location within the Brown's Pond study area is characterized according to the percentage amount of potential solar radiation it is likely to absorb at a given point in time. Note that each shading pattern represents not just one zone but a range of percentages.

Once local surface characteristics have been detected and expressed in the form of values associated with individual locations, it remains to aggregate these locational values over surficial conditions. In the case of both lineal and areal characteristics, this is generally a matter of summarizing values within zones. In the case of surficial characteristics, however, it is usually a matter of summarizing values within extended neighborhoods.

Consider, for example, a surficial analogue to the roundness-measuring technique presented in Fig. 7-6. There, a value was computed for each location within a given zone by dividing the square root of the zone's area by its perimeter. Here, a *surficial roundness* value would be computed for each location on a surface by dividing the cube root of surface volume by the square root of surface area within a specified vicinity. If that ratio is then multiplied by 220, it will yield a figure ranging from a maximum of approximately 100 for a hemisphere to a minimum approaching zero for highly undulated forms. This technique might be applied to an *Altitude* layer, for example, as follows:

SurfaceVolume	= *IncrementalVolume of Altitude*
SurfaceVolume	= *FocalSum of SurfaceVolume at ... 100*
SurfaceVolume	= *LocalRoot of SurfaceVolume and 3*
SurfaceVolume	= *LocalProduct of SurfaceVolume and 220*
SurfaceArea	= *IncrementalArea on Altitude*
SurfaceArea	= *FocalSum of SurfaceArea at ... 100*
SurfaceArea	= *LocalRoot of SurfaceArea and 2*
HowSpherical	= *LocalRatio of SurfaceVolume and SurfaceArea*

A similar technique is illustrated in Fig. 7-14. Here, each location's value on the *SmoothAltitude* layer shown in Fig. 5-16 has been subtracted from its value on the *Altitude* layer shown in Fig. 1-9 as follows:

Deviation	= *LocalDifference of Altitude and SmoothAltitude*
YinYang	= *LocalRating of Deviation*
	with 2 for ... -11 with 3 for -10 ... 10 with 1 for 11 ...

The resulting *YinYang* layer indicates the difference between each location's elevation and a distance-weighted average of the elevations nearby. Note that this result is analogous to that of the areal protrusion-measuring technique presented in Fig. 7-10. Here, positive and negative deviations in surficial elevation values serve to define areas of *surficial intrusion* and *protrusion*. Once identified as such, the planar "footprints" of these hills and valleys can then be processed as areal conditions with a whole new set of shape and size characteristics.

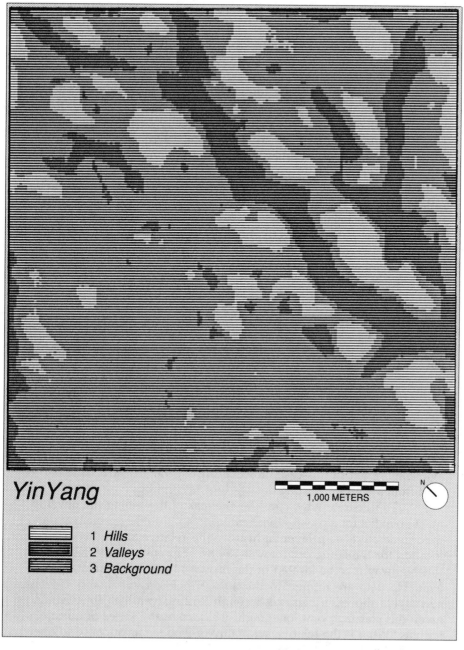

Figure 7-14 A map layer indicating surficial protrusion and intrusions. *YinYang* is a map layer depicting negative (yin) and positive (yang) deviations from the general trend of the topography in the Brown's Pond study area.

7-3   SYNTHESIS OF CARTOGRAPHIC CHARACTERISTICS

For a number of descriptive cartographic modeling applications, analytic techniques like those presented in the previous two sections will suffice. This is particularly true of scientific applications intended to generate objective knowledge. Applications in areas such as land use planning and environmental management, on the other hand, tend to involve a more subjective relationship between the user of geographic data and that which the data represent. Here, descriptive cartographic modeling techniques are used not only to expose the significant facts embodied in a set of data, but also to express the meaning a user may attribute to those facts. Such techniques are synthetic rather than analytic in nature. They attempt to synthesize information from data using judgment.

To be able to effectively synthesize a wide range of cartographic characteristics, it is useful to adopt certain fundamental synthesis techniques. These techniques can be expressed in terms of the
- formulation and
- implementation

of descriptive cartographic models.

Formulating a Descriptive Model

It may seem that the obvious place to begin formulating a descriptive cartographic model is with the initial set of data from which the model will be constructed. If these data have already been collected and if no additional data are likely to be forthcoming, it may seem especially prudent to start with what resources are actually available. Even in this case, however, it is generally advisable *not* to start from a body of existing data.

To illustrate why, consider the task of describing the development potential for a prospective land use. Suppose the problem at hand is to generate a map layer indicating the relative suitability of each location for the siting of a new highway. Suppose, furthermore, that this highway is to be located in the Brown's Pond study area and that available data are limited to the *Altitude*, *Water*, *Vegetation*, and *Development* layers shown in Figs. 1-9 through 1-12. Given that objective and given this starting point, one might well arrive at a series of questions such as "What is the relative suitability of wooded as opposed to open land for the siting of this new highway?"

In attempting to answer such a question, it quickly becomes clear that this is not a simple matter. Issues ranging from the economic cost of clearing trees and the aesthetic qualities of vegetation to the potential impact on wildlife habitat and concern for snow removal must somehow be integrated before any such judgment can be expressed. And this would have to be done for each of the zones on each of the layers of data felt to be pertinent.

In doing this, however, it would probably soon become apparent that certain types of siting criteria often tend to recur. The issues of construction cost, user appeal, environmental impact, and maintenance that arise with respect to vegetation, for example, may also arise with respect to topographic, hydrographic, or demographic conditions.

With this in mind, it is often helpful to try to anticipate such general issues before interpreting specific data. A reasonable way to begin to do this is by listing criteria without immediate concern for the data from which they will ultimately be derived. Ideally, such a list would exhaust all conceivable siting criteria in a clear and consistent manner. More typically, however, such a list will be incomplete and will include criteria defined at different levels of abstraction. A list of highway siting criteria, for example, might include concerns ranging from "economic viability" and "environmental quality" to "that wonderful view of Mount Wachusett at sunrise in early October."

One way to improve on a list of criteria in which some items are quite general in nature while others are much more specific is to reorganize those items into a hierarchical order. That early morning vision of Mount Wachusett, for example, might be cited as one of several issues relating to visual quality. This, in turn, might be grouped with concerns such as aural or tactile quality under the rubric of aesthetics. And as more and more general groupings are formed, aesthetics might well appear as a component of environmental quality before this is combined with similar abstractions into an index of site suitability.

The process of reorganizing a random or sequential list of siting criteria into a hierarchical ordering can often help not only to improve clarity and consistency but also to identify significant criteria that might otherwise be overlooked. This is typically a matter of induction and deduction to find the "parent" of a given issue and then to infer its "offspring." Having identified that concern for the view of Mount Wachusett, for example, one might induce that this is a matter of visual quality and then go on to deduce that certain other views should be considered as well. The attempt to define a geographic characteristic at multiple levels of abstraction will often afford a better understanding of the characteristic itself.

Rather than listing and then reorganizing a set of siting criteria, it may be more efficient to think in terms of a hierarchy from the outset. We have already suggested that this can be done by working either inductively (from particular instances to general principles) or deductively (from general principles to particular instances). We have also suggested, however, that a purely inductive strategy can become both tedious and nearsighted.

The deductive approach is generally much better. By starting not with an inventory of available data but with a conception of the intended output, a "divide and conquer" strategy can be used to break otherwise formidable tasks into manageable pieces. Complex issues can be expressed in terms that need not be fully defined right away but, instead, are presented as abstractions. These can then be defined more and more specifically as the composite characteristic that must ultimately be described is decomposed into finer and finer components.

At each level in this hierarchy, the subcomponents of a given component should ideally satisfy three conditions. They should be

- all-inclusive,
- mutually exclusive, and
- meaningful.

The first of these conditions demands that subcomponents collectively account for all pertinent aspects of the component they serve to define. A distinction drawn between "past, present, and future" or "costs and benefits," for example, would almost certainly satisfy this condition, while a breakdown into "average and below average" would leave a good bit unaccounted for.

The second condition demands that each component encompass its own distinct set of concerns, such that no one issue is addressed as part of more than one higher-level component. Overlapping categories such as "red, white, blue, and *dark*" can lead to redundancy and confusion. In practice, however, it is sometimes difficult to maintain clear distinctions. Suppose, for example, that the siting criteria for a proposed land use were to involve three major factors including steepness, wetness, and openness. The ideal site is flat, dry, and wooded. Now consider the site suitability of a lake or a river in terms of these factors. When assessing steepness, those bodies of water should be designated as flat and therefore suitable. Their apparent wetness and openness should not be of concern at this point but should be addressed only when each of those two factors is considered in its own right. By keeping issues clearly separate until they are finally combined, the combinatorial process itself can be much more precisely and flexibly controlled.

The third condition for successful decomposition of cartographic characteristics into hierarchical components demands simply that these components be of useful significance. Suppose, for example, that a set of concerns were to be classified into 26 groups respectively including those whose names begin with the letter *A*, those beginning with *B*, with *C*, and so on through *X*, *Y*, and *Z*. While these groupings would certainly provide all-inclusive and mutually exclusive coverage, it is rather unlikely that they would prove to be meaningful components. A meaningful component will generally represent a set of phenomena that either acts or can be acted upon as a whole.

This hierarchical decomposition or stepwise refinement technique is a widely used strategy not unlike the *outlining* approach to writing or the *top-down* approach to software design. In practice, it is a technique that inevitably involves a good bit of inductive as well as deductive thinking. It is also a technique that inevitably involves an ongoing struggle between conceptual elegance and practical utility. Loose ends and leaps in abstraction must often be held in abeyance, and issues must often be simplified in order to be managed at all. As an ideal, however, this deductive approach to the formulation of descriptive cartographic models is well worth its sometimes difficult pursuit.

Implementing a Descriptive Model

The branches and twigs of a cartographic model constructed in this tree-like form will ultimately terminate in leaves corresponding to available or acquirable data. It is significant that these leaves will have been propagated from roots that can be equated with the model's anticipated products. As such, it is these products that will dictate both the organization and the content of the cartographic model's initial data base rather than vice versa.

This can also help to reduce the often substantial amounts of time and energy associated with geographic data preparation. By dissecting information needs into elementary components, it may well be discovered that certain components can be generated from others. By noting similarities among these components, they can often be combined or generalized into a more efficient form. And by estimating the degree to which each component is likely to be used, those of least utility can sometimes be justifiably eliminated.

Once the leaves of the conceptual tree representing a descriptive cartographic model have been expressed in the form of existing map layers, the model can then be implemented in the form of a procedure. To do so, the branches of this tree must be pruned one-by-one from the leaves back to the roots. Pruning in this context is a matter of specifying the operation or operations necessary to generate each new or intermediate map layer from one or more of the map layers initially encoded or previously derived.

Each of these generated layers should ideally correspond to one of the hierarchical components cited in the initial formulation of the cartographic model. In practice, however, this can be difficult. It is not always clear how an issue such as ecological diversity or historic significance can be expressed in the form of numerical values assigned to specific locations. One technique is to rely on measurements of readily observable characteristics that can be used as indicators of qualities that are otherwise difficult to measure. Predominant vegetation type, for example, might be used as an indicator of wildlife habitat quality. Or the number of recently sold homes within a half-mile radius might be used as an indicator of neighborhood instability.

As the conceptual plan of a cartographic model is translated into the more tangible form of map layers and operations, subjective elements of the model must also be expressed in concrete terms. In contrast to the kind of subjective judgment that is implicitly involved in formulating the structure of a descriptive model, subjective assertions embodied in the substance of a model must be explicit.

This can be facilitated by separating those portions of the cartographic model that are primarily objective and analytic in nature from those that are more subjective and synthetic. Note that the former tend to be concerned with questions of what is significant, while the latter are concerned with how it is significant and how significant it is. By separating the two, descriptions of *what* can be expressed in terms of standard observations and measurements without being complicated by more specialized interpretations of *how*. Separating the two may also facilitate teamwork, since objective analyses are likely to be conducted by technical specialists and are likely to produce definitive results, while subjective syntheses are more likely to be undertaken by generalists and are more likely to be subject to revision.

To achieve this separation of analytic and synthetic components in a cartographic model, each component must be able to communicate with others. This, of course, is a matter of using map layers as common media. Communication can also be facilitated by standardizing the way in which these layers are combined. It is generally advisable to integrate the components of a cartographic model with location- rather that neighborhood- or zone-characterizing operations. In this way, a wide variety of combinatorial methods can be expressed in one clear and consistent manner. All will involve some form of *LocalFUNCTION* operation.

This use of *LocalFUNCTION* operations will generally involve an application of subjective judgment to objective data in either of both of two ways. It may occur when the individual components of a model are defined, or it may occur when these components are combined.

In the first of these cases, the expression of subjective judgment is typically just a matter of using *LocalRating* to characterize a layer's zones in terms of their implications. The major decision at this point is whether to express these implications in terms of nominal, ordinal, interval, or ratio scales of measurement. It is generally a good idea to be as quantitative as possible, but only when such quantification can be supported with meaningful units.

While the synthesis of a cartographic characteristic may involve no more than a single application of the *LocalRating* operation, it will usually involve several such applications. These will generate a set of layers that must ultimately be combined. In this second type of situation where subjective judgment is to be applied, methods will also depend on the nominal, ordinal, interval, or ratio nature of the values to be combined.

Perhaps the most widely applicable way of combining values is with *LocalCombination*. This operation can be used to synthesize all types of values and to do so in a way that avoids (or at least defers) any precise expression of judgment. The only major problem with this approach is that *LocalCombination* output is not readily transferrable. Since the value it assigns to any particular combination may vary from one study area to another, this operation is of limited use in building generic models.

The next most widely applicable way of synthesizing values is by applying the *LocalRating* operation to two or more map layers. Nominal, ordinal, interval, and ratio values can all be combined in this manner, and combinations can be characterized without reference to any one study area. To do so, however, not only demands an expression of judgment but an expression that must be explicitly applied to all possible combinations.

The alternative to this is to synthesize values in the form of a common statistic. To do so, all values must relate to similar units of measurement, and the statistic involved must be appropriate for use with this measurement scale. Values relating to any scale of measurement can be synthesized using the *LocalMajority*, *LocalMinority*, or *LocalVariety* functions, while values of at least ordinal significance can also be combined with *LocalMaximum* and *LocalMinimum*. Interval-scale values can be summarized with *LocalSum*, *LocalDifference*, and *LocalMean* as well, while *LocalProduct*, *LocalRatio*, and *LocalRoot* are generally appropriate for use only with values on a ratio scale.

As indicated earlier, the particular choice of statistic by which to combine a set of values will also depend on whether those values are to be represented in terms of their variation, a typical case, an atypical instance.

7-4 QUESTIONS

This chapter has suggested techniques by which individual cartographic modeling capabilities can be combined to build descriptive models of a variety of geographic phenomena. To further examine these techniques, consider the following questions.

7-1 In Chapter 1, it was asserted that "raster structures are generally better suited to the interpretation of *where*, while vector structures are better suited to the interpretation of *what*." How does this assertion relate to the distinction between interpretive operations respectively associated with individual locations, locations within neighborhoods, and locations within zones?

7-2 A layer of roads and a layer of streams are combined into a layer of probable bridge locations. Is the operation involved analytic or synthetic in nature?

7-3 Suitability for the siting of a domestic sewage disposal system varies according to five major factors: soil permeability, soil depth to groundwater, soil depth to bedrock, proximity to surface water, and topographic slope. Suppose that each of these factors is depicted on a different map layer with values ranging from one (least suitable) to 10 (most suitable). How should these layers be combined into a single layer of site suitability?

7-4 How would you generate a layer on which the upper left quadrant of the Brown's Pond *Water* layer is effectively magnified by a factor of two, such that each 20-meter grid square is now represented by four 10-meter grid squares? How could you then smooth the resulting "stair-step" form of boundaries? How would this process differ for the Brown's Pond *Altitude* layer?

7-5 How would the results of the hole-counting technique illustrated in Fig. 7-8 be likely to differ if applied to something other than a layer generated by *FocalInsularity*? How could you count the number of different fragments that comprise a zone?

7-6 Why does the *InferredMobility* layer shown in Fig. 7-1 differ at all from the *Mobility* layer shown in Fig. 4-2? How could the *InferredMobility* shown in Fig. 7-1 be "cleaned up" by filling in and smoothing off what appear to be lineal forms, while erasing that which does not appear to be part of this lineal network?

7-7 How can the minimum-cost path be extracted from the *BestTime* layer shown in Fig. 7-5?

7-8 How would the technique used to generate the *BestPath* layer shown in Fig. 7-3 have to be modified if both path termini were areas made up of multiple locations? How would it have to be modified to generate a second best path?

7-9 How could you generate a layer on which road segments that are generally oriented along north-south axes are distinguished from those with east-west orientations? How about distinguishing straight roads from those that are curved?

7-10 A technique used in image processing to sharpen black-and-white photographs is to transform transitions from dark to light into transitions from dark to very dark to very light to light. Given a photograph in the form of a map layer on which higher values represent lighter shades of gray, how could this be done using cartographic modeling operations?

7-11 "It's that hill just south of Brown's Pond and southeast of Petersham Center, the one that is almost surrounded by streams." This "simple" description of spatial relationships should be easy to interpret with the power of a geographic information system. Right? How about "the first pond on your left as you head south from the town commons on Main Street"?

7-12 One of the important topographic forms in ancient Chinese geomancy is the serpentine or "dragon-shaped" hill. How could such a form be detected on a surface of topographic elevation values?

Chapter 8

PRESCRIPTIVE MODELING

Given the ability to analyze and to synthesize cartographic data, a variety of cartographic models can be developed to represent facts, to simulate processes, to express judgment, or to otherwise provide for effective description of geographic phenomena. To move from description to prescription, however, a new set of techniques is required. These are techniques that broaden the role of cartographic modeling from relatively passive inquiry to much more active intent. Descriptive models answer questions; prescriptive models solve problems.

The problems addressed by prescriptive cartographic models generally involve some form of *cartographic allocation*, the process of selecting locations in order to satisfy stated objectives. A cartographic allocation problem may range from the siting of a proposed land use to the routing of a freight shipment or the reapportionment of political districts to the choice of a place to eat. This sort of decision making is often done entirely "by eye" and is seldom fully automated. Nonetheless, computation can greatly enhance our human ability to generate and evaluate alternative solutions to problems.

While the purpose of an allocation model is ultimately to prescribe, the model itself will typically encompass both descriptive and prescriptive parts. These are typically embodied in three major steps
- the initial statement of a problem,
- the generation of a solution to this problem, and
- an evaluation of results.

The statement of an allocation problem is a descriptive task. In a cartographic modeling context, this will typically begin with an explicit specification of some geographic quality to be achieved. It will then include an equally explicit description of how this quality arises from geographic conditions that either exist or could be made to exist on a given site. For example, if minimal soil erosion is a quality to be achieved, the erosion process must first be described in terms of cause and effect. The amount of soil that is likely to be lost from a given amount of land over a given amount of time must be expressed in terms of factors such as rainfall, soil type, topography, and land use. This can be done by way of descriptive cartographic modeling techniques.

To move from the description of a problem to a prescription for its solution generally requires that the descriptive model of the problem be *inverted*. The descriptive model will describe variation in a particular geographic quality as a function of existing site conditions and potential site modifications. Its prescriptive counterpart must express a range of potential site modifications as a function of existing site conditions and the geographic quality sought. In the case of the soil erosion model, for example, the one factor likely to be subject to modification is not rainfall, soil type, or topography, but prospective land use. To determine the land use or uses (perhaps including alternative types of agricultural practice) that result in acceptable erosion levels, a model must be developed that expresses land use possibilities as a function of rainfall, soil type, topography, and allowable erosion levels.

If this were to be represented by way of conventional rather than a cartographic algebra, the initial description of erosion might be presented in the form of an equation such as

$$E = R * S * T * L$$

where E is an estimated level of erosion and R, S, T, and L are multiplicative factors respectively associated with rainfall, soil type, topography, and land use. Inversion of this descriptive model into prescriptive form would then simply be a matter of algebraic transformation to yield the equation

$$L = E \ / \ (R * S * T)$$

Note that this expression indicates the value of the land use factor L that will be necessary to achieve an acceptable level of erosion E. As such, it prescribes a solution to the problem initially stated.

Similar analogues to the process of inverting a descriptive cartographic model into prescriptive form can be seen in the use of symbolic logic, in the application of mathematical optimization methods, and even in the most casual forms of personal decision making. In each case, however, inversion is accomplished through a different set of techniques.

In the case of cartographic modeling, these techniques will generally yield a solution in the form of a map layer on which locations that have been selected are associated with zones corresponding to recommended site treatments. A solution to the problem of limiting erosion, for example, might be presented as a layer on which each zone corresponds to a recommended land use.

Once an optimal or at least satisfactory solution to a problem has emerged, it is often useful to evaluate just how well the problem has been solved. This is particularly true when a problem has been cast in terms of multiple and potentially conflicting objectives or especially severe constraints. Evaluation is generally a matter of comparing the quality(ies) actually achieved by a proposed solution to those predicted by the descriptive model on which that solution was based. In the case of the soil erosion model, this would involve a comparison of predicted erosion levels with and without the proposed land use.

There are a variety of cartographic modeling techniques that can be used to express problems, to derive solutions, and to assess results as part of an exercise in cartographic allocation. One way to classify these techniques is to draw a broad distinction between
- *atomistic* allocation and
- *holistic* allocation.

Atomistic allocation problems are those that can be expressed in terms of individual pieces or "atoms" of geographic space, while **holistic** problems cannot be expressed except in terms of geographic "wholes." Allocation problems can also be classified according to the number of zones to be allocated and the number of criteria to be satisfied.

8-1   ATOMISTIC ALLOCATION

From a cartographic modeling perspective, atoms of space exist in the form of individual locations. An atomistic allocation problem is therefore one that can be addressed on a location-by-location basis. As with any form of allocation, this will require that we
- state the problem,
- generate solutions, and
- evaluate results.

Stating the Problem

As indicated earlier, the statement of a cartographic alloca- tion problem is a matter of expressing some desired geographic quality as a function of existing site conditions that are presumed to be constant and prospective site conditions that are subject to control. In the case of atomistic allocation, this can be done in a straightforward manner. To do so requires only that a cartographic model be constructed in which map layers of all pertinent site conditions are combined using *LocalFUNCTION* operations to generate a new layer on which each loca- tion is characterized in terms of the quality sought.

To express the problem of limiting soil erosion in terms of atomistic allocation, for example, we might begin with map layers of rainfall, soil erodibility, topographic steepness, length of slope, and any other existing site condition felt to be significant. These would then be combined with a layer showing all potential sites (typically including the entire study area) for a proposed land use such as live- stock grazing. The result of this combination would be a new map layer predicting rates of erosion due to grazing.

To complete the problem statement, we must describe the level of desired environmental quality to be achieved. This can generally be done in either of two ways: by stating that the quality achieved must be above a specified level, or by stating that it must simply be as high as possible. In the case of soil erosion due to livestock grazing, for ex- ample, we might establish a maximum limit on tons of lost soil per acre per year. Or, we might simply assert that erosion levels should be minimized.

With this kind of problem statement, we effectively establish a basis for the allocation of a single cartographic zone in response to a single criterion. We can easily extend this form of problem statement, however, to also address allocation problems involving multiple zones and/or multiple criteria.

To allocate a single zone in response to multiple criteria, each criterion must first be expressed in the manner described above. In the case of livestock grazing, for example, the same kind of map layer that was generated to estimate levels of soil erosion might also be generated to estimate levels of aesthetic quality, site development cost, wildlife habitat value, and so on. Each of these layers would also have to be accompanied by a statement of quality requirements.

And if more than one zone is to be allocated, the whole process must be repeated for each zone involved. A multiple-zone allocation problem might call for the siting of land uses such as housing, mining, or conservation, for example, in addition to livestock grazing. Though each of these land uses will have its own set of siting criteria, it will often be possible to relate these criteria to a common set of issues. In this way, the multiple-zone allocation problem can be structured as shown in Fig. 8-1. Note that this structure makes it possible to summarize criteria either by issue or by zone.

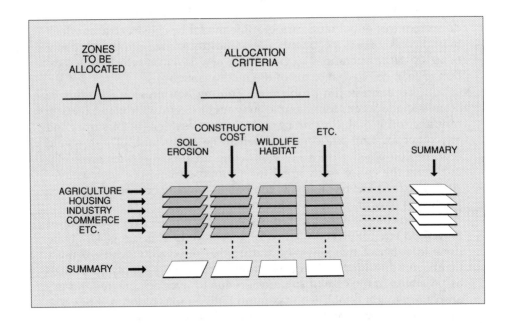

Figure 8-1 Organization of an atomistic allocation problem involving multiple zones and multiple criteria. By organizing a multiple-zone allocation problem such that the siting criteria for all zones relate to the same set of issues, those criteria can be expressed in the form of map layers as shown. Each of the gray layers depicts the level of a particular environmental quality that would be achieved at each location if that location were to be occupied by a particular type of zone. The white layers summarize these qualities by zone or by criterion.

Generating a Solution

Given a map layer indicating the amount of soil erosion likely to result from livestock grazing, and given an objective stating that such erosion should be limited or minimized, the task of inverting this descriptive problem statement into a prescriptive problem solution is simply a matter of selecting those locations with acceptable erosion estimates. The process can become more complicated, however, when conditions in a particular study area are such that the allocation problem proves to be over- or under-constrained.

If a problem is over-constrained, constraints must be relaxed and/or objectives made less demanding. In the case of the erosion problem, an over-constraining situation would be one in which there is no location at which any prospective land use would result in an acceptable level of erosion. Here, since the constraints involved are a matter of physical reality and not a matter of discretion, the stated objectives of the allocation problem would have to be revised. This might involve increasing the level of erosion deemed acceptable or simply stating that erosion should be kept as low as possible.

When an allocation problem is under-constrained, a solution can be generated by tightening constraints, heightening objectives, or simply selecting arbitrarily from among available choices. In the case of the erosion problem, an under-constraining situation would be one in which there are more acceptable locations than needed to site prospective land uses. Rather than tamper with constraints or allocate arbitrarily in this situation, it would probably be best to make the objective of the problem more demanding. One way to do this would be to place stricter limits on allowable erosion. Another would be to require that erosion levels be kept as low as possible (ironically one of the same strategies that can be used to soften objectives when a problem is over-constrained).

When an atomistic allocation problem calls for the siting of a zone in response to more than one criterion, a solution can be generated by selecting locations from a layer that combines the quality predictions for all criteria involved. Suppose, for example, that livestock grazing areas were to be sited considering soil productivity and availability of water as well as erosion control. Each of these criteria would first have to be represented in the form of a map layer indicating how well livestock grazing would fare in terms of that one criterion. The three layers would then have to be combined to indicate how well grazing would fare in terms of some composite measure.

If this were done using the *LocalMean* operation, it would suggest that a strong performance on one criterion could compensate for a weak performance on another. If *LocalMaximum* were to be used (with maximum value indicating maximum quality), it would indicate that a strong performance on any one criterion is sufficient. If *LocalMinimum* were to be used, however, this would indicate that the most significant criterion is whichever is satisfied least.

When an atomistic allocation problem calls for the siting of more than one zone, each zone can be allocated in the manner described above. The only major problem with this approach is that two or more zones may well be allocated to a common set of locations. To resolve such conflicts, we must somehow relate the siting of each zone to the siting of all other zones.

One way to do this is to allocate zones in a predefined order such that the first zone assigned to a given location is the only zone that can be assigned to that location. The order of allocation may be designed to favor those zones that will yield the greatest benefit if well sited, those zones that will cause the greatest problems if poorly sited, those zones that have the most restrictive siting requirements, those zones that will have the greatest influence on the siting of other zones, and so on.

Another way to coordinate the siting of multiple zones is to seek out those locations that are well suited for one zone and not well suited for any others. Not only does this help to avoid conflicts, it also tends to maintain options for future siting decisions by keeping available those locations with the broadest development potential.

Evaluating Results

Once a set of locations has been allocated in response to atomistic criteria and recorded as a zone on a new map layer, the degree to which these locations satisfy those criteria can be easily assessed. To do so requires only that the new map layer be compared to the composite layer or any of the component layers in the descriptive model initially used to express those criteria. In the soil erosion case, for example, the layer showing locations selected for livestock grazing might be compared to the layer of erosion estimates. This would indicate where and how much erosion is likely to occur as a result of livestock grazing.

If two or more criteria were used in allocating a zone, then this process would be repeated for each criterion involved. And if multiple zones were allocated, the evaluation process would have to be conducted such that each zone is assessed in terms of its own set of criteria. The structure presented in Fig. 8-1 can facilitate this process.

8-2 HOLISTIC ALLOCATION

Holistic allocation problems are those that cannot be addressed by considering locations one at a time but only by treating groups of locations as integrated wholes. To illustrate this, consider again the problem of limiting the amount of soil erosion associated with new livestock grazing areas. Though it was never formally included as part of the original problem statement, it was nonetheless implied (and very probably assumed) that this problem called for a certain minimum amount of grazing land to be allocated. If that were not the case, then this allocation problem could be regarded as truly atomistic. By incorporating the size consideration, however, we introduce a quality that is not manifest in any one location. It arises only when multiple locations are treated as a whole.

This would be true to an even greater extent if we were to require that allocated locations be conterminous or that they collectively form a particular shape. Suppose, for example, that a criterion for the siting of a new land development project were to stipulate that it occupy exactly 918 locations in the shape of the letter S. Now consider the kind of decisions that might be applied to a candidate location. Suppose its soil type is good, its purchase price is fair, and all other pertinent site conditions are found to be acceptable. Still, that location must ultimately join with 917 other locations to form an S. Is it a good location or not?

While this problem may seem contrived, it is not unlike the problem of siting an airport with a particular configuration of runways, the problem of routing a highway to connect two locations at the lowest possible cost, or the problem of designing a wildlife preserve to encompass several different types of landscape. In each of these cases, allocation criteria are holistic, and they call for a somewhat more sophisticated set of prescriptive techniques. Like their atomistic counterparts, however, these techniques can still be expressed in terms of three major steps to
- state the problem,
- generate a solution, and
- evaluate results.

Stating the Problem

An holistic allocation problem statement must describe the relationship between existing site conditions that are presumed to be constant, prospective site conditions that are subject to control, and selected environmental qualities that arise from these conditions. To that extent, it is similar to its atomistic counterpart. Both types of allocation problem can in fact be stated in a similar manner: by constructing a cartographic model that generates a map layer of the desired quality from layers depicting both existing and proposed site conditions. What distinguishes the holistic model is its depiction of the site conditions that do not yet exist.

To describe holistic qualities arising from existing conditions is generally a matter of applying neighborhood- and/or zone-characterizing capabilities to layers depicting those conditions. The steepness and slope length factors in the soil erosion model, for example, are holistic qualities arising from existing topographic conditions. Once these qualities have been measured and recorded in the form of explicit values associated with individual locations, however, they are no longer holistic in nature. They have effectively been *atomized*.

Holistic qualities that arise from not-yet-existing site conditions can also be described by applying neighborhood- and/or zone-characterizing capabilities. But to what map layer(s) are these capabilities to be applied? In the atomistic case, we could present a layer depicting all potential locations (typically including the entire study area) for a proposed land use or other prospective site condition. In the holistic case, however, we need something much more specific. Before we can measure holistic qualities relating to characteristics such as shape or size, we must have a clear indication of which particular locations contribute to that shape or size and which do not.

It is a chicken-and-egg dilemma. Holistic qualities arising from prospective site conditions cannot be measured until those conditions have been allocated, but site conditions cannot be allocated without first measuring those qualities.

At this point, all we can do is to construct the descriptive model by referring to a not-yet-generated layer of selected locations. To actually implement the model, we must first generate a solution.

Generating a Solution

Even though the descriptive model that is used to state an holistic allocation problem cannot be implemented until a solution to the problem has already been generated, the process of generating such a solution is still one of inverting the descriptive model into prescriptive form. In the case of atomistic allocation, inversion is facilitated by the fact that causes and effects are expressed such that they occur at common locations. Though this localization of cause (existing and proposed site conditions) and effect (resulting environmental qualities) is more difficult to achieve in the holistic case, it is no less important in facilitating the process of inversion.

To generate solutions to holistic allocation problems generally calls for techniques that include and refine those associated with atomistic allocation. The refinements are only minor in some cases but quite substantial in others. In any case, it is generally best to begin by isolating those portions of the descriptive model that involve holistic qualities and then to atomize these as much as possible.

One of the simplest and most common types of holistic allocation problem has already been mentioned. It is the one in which siting criteria that are otherwise entirely atomistic in nature include an holistic requirement specifying the number of locations to be selected. The solution strategy in this case is straightforward. First, the atomistic portion of the problem's descriptive model must be run in order to estimate quality levels at individual locations. Given this information (and assuming that the sum of these quality levels is to be maximized), allocation is simply a matter of starting with the location(s) of highest quality and adding others in order of decreasing quality until the total number of locations required has been met.

For all its simplicity, this strategy embodies a general approach that can also be used to address increasingly sophisticated problems. The approach involves two steps. First, a map layer is created to indicate the level of quality estimated for each location within a study area as a function of atomistic conditions. Next, the set of holistic siting criteria are applied.

This general approach can also be seen in a slightly more complex example. Suppose that locations are to be selected such that not only do they satisfy a set of atomistic criteria and a specified size requirement, but also a requirement stating that their spatial pattern should be "sparse." Such a requirement might be imposed in siting vacation homes, for example, or competing business establishments.

To effectively respond to this holistic criterion, we must attempt to atomize it. We must try to determine, for each location that will ultimately be selected, the number of other locations within its vicinity that will also be selected. A careful reading of this task (and a healthy suspicion of chickens and eggs) will reveal that it cannot be accomplished as explicitly stated. What can be done, however, is to determine the number of locations that are *likely* to be selected in the vicinity of any given location. To do this, we need only apply an operation such as *FocalMean* to the layer of quality levels resulting from atomistic conditions. The result will be a layer on which each location is characterized according to the overall attractiveness of atomistic conditions in its neighborhood.

This measure can then be used as an indicator of sparsity by accepting an assumption that is speculative but nonetheless critical. The assumption is that an attractive neighborhood will be less conducive to sparse development. If we can accept this assumption, we can treat a location's proclivity toward isolation from allocated neighbors (hence sparsity) as another of its atomistic site conditions.

An example of this technique is shown in Fig. 8-2. Here, a layer entitled *NeighborScore* has been created from the *LowScore* layer shown in Fig. 4-15 as follows:

NeighborScore   = *LocalRating of LowScore with -0 for 0 10 with 1 for 20 30*
NeighborScore   = *FocalSum of NeighborScore at ... 1000*

Note that locations with *NeighborScore* values of from one to 20 are those most suitable for sparse development.

What makes this technique speculative is its reliance on a *heuristic* rather than an *algorithm*. Heuristics and algorithms can both be defined in general terms as sets of directions to be followed in order to satisfy stated objectives. An algorithm gives directions expressed in terms of specific responses to conditions that are presumed to be known with certainty. The directions given by a heuristic, on the other hand, are expressed in terms of responses that may or may not be specific. Furthermore, these are responses to conditions that (though always possible and usually probable) are not necessarily certain. An algorithm will chart a definitive path all the way to a predictable outcome. A heuristic will merely offer advice for use along the way toward an outcome that is likely but never fully guaranteed. Thus, the heuristic is not so much a set of guideposts by which a problem can be solved as it is a set of guidelines by which a problem can be explored.

This exploratory process is typically one in which a series of steps are taken (or a given step is repeated) such that each step attempts not only to solve a part of the problem but also to provide new information for whatever steps may follow. In allocating the sparse development pattern, for example, we employ a heuristic asserting that isolated locations with good site conditions should be selected over clustered locations with equally good conditions. If we apply this logic repeatedly, its first application will identify those locations that can be selected (since they are isolated) without fear of jeopardizing the ultimate goal of a sparse pattern. Once these locations have been selected, they can be treated as "existing" conditions themselves and thereby used to reduce uncertainty when the same step is applied to

Figure 8-2 A map layer created by atomizing an holistic characteristic. *NeighborScore* is a map layer on which each location within the Brown's Pond study area that is suitable for housing development is set to a value indicating the number of equally suitable locations within 1,000 meters. Note that each shading pattern represents not just one zone but a range of values.

remaining locations. Ultimately, this process will lead to a solution that may not be demonstrably best but does at least strive for that goal and will probably be quite good.

The allocation strategy employed here can also be used to achieve other holistic qualities associated with spatial form. To achieve density rather than sparsity in an allocated pattern, for example, we need only modify the heuristic slightly. Instead of favoring those locations that have good conditions themselves and relatively bad conditions nearby, we would favor those whose good conditions are also shared by neighbors.

As the holistic qualities to be achieved in a cartographic allocation problem are defined in more specific terms, solution strategies tend to become more complex. Consider, for example, the effect of a siting criterion requiring that allocated locations be conterminous. One way to achieve a pattern of locations that are all contained within a single boundary would be to apply the dense housing pattern allocation technique mentioned above. This will tend to allocate conterminous forms, but only those that are also quite compact.

To allocate forms that are conterminous without necessarily being compact, another approach is to generate a layer on which all locations whose atomistic conditions are deemed acceptable are aggregated into conterminous groups with *FocalInsularity*. An example of this technique is presented in Fig. 8-3. Here,

SiteSize	=	*LocalRating of LowScore with -0 for 0 10 with 1 for 20 30*
SiteSize	=	*FocalInsularity of SiteSize*
SiteSize	=	*IncrementalArea of SiteSize*
SiteSize	=	*ZonalSum of SiteSize*
SiteSize	=	*LocalRating of SiteSize with 1 for ... 36000 with 2 for 36001 ...*

has been applied to the *LowScore* layer shown in Fig. 4-15 to generate a layer distinguishing between large and small conterminous parcels of acceptable locations. The drawback to this approach is its requirement that "acceptable" atomistic conditions be specifically predefined. Since this must be done before the holistic criterion is ever considered, there is no opportunity to make trade-offs between the two.

This can be remedied by way of a third technique for conterminous pattern allocation that combines characteristics of the previous two. Suppose we were to apply the dense-pattern allocation technique mentioned above with a *spreading in FRICTIONLAYER* version of the *Focal-Sum* operation. Suppose, furthermore, that this were to be done with a *FRICTIONLAYER* on which the locations with lower travel costs are those with the better atomistic conditions. In this way, the neighborhood of

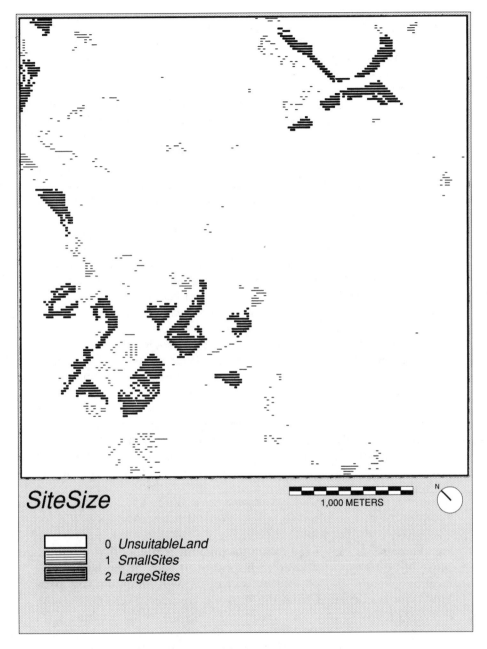

Figure 8-3   An allocation of conterminous housing sites.   *SiteSize* is a map layer identifying those locations within the Brown's Pond study area that are individually suitable for
housing and collectively form conterminous parcels that are either above or below a
specified size.

atomistic conditions to be examined around each prospective location will become more and more skewed toward those atomistic conditions that are most suitable.

This technique embodies a strategy similar to that of the optimal path-generating procedure illustrated in Fig. 7-3. The similarity should not be surprising, since both techniques are intended to allocate a set of locations that not only satisfy atomistic conditions but that also form a conterminous pattern. In the case of the path allocation problem, we have merely added a requirement that this form connect two locations. By adding that requirement, we also manage to formulate one of the few types of cartographic allocation problem that *can* be solved by algorithm. As we extend the problem, however, we must return to heuristic methods.

Consider, for example, the problem of laying out logging roads for a timber-harvesting operation. Given a destination such as a sawmill or perhaps the nearest paved highway, this problem is one of determining which pattern of roads provides the most efficient overall access to harvestable acres at the lowest overall construction cost. A network of minimum-cost paths must be generated to provide efficient access from the destination location(s) to many others.

One way of beginning to attack this problem is illustrated in Figs. 8-4 through 8-6. Fig. 8-4 presents a layer entitled *InitialTransportCost* on which each location has been characterized in terms of the (purely hypothetical) cost of constructing a segment of logging road across it. Among the factors used to establish this cost were existing development, water bodies, and topographic slope. In Fig. 8-5 is a layer that uses these costs as measures of distance from each location to a major highway. This *InitialAccessCost* layer was created from *InitialTransportCost* and the *Development* layer shown in Fig. 1-12 as follows:

BigRoad   = *LocalRating of Development with -0 for 0 2 3 4 5*
InitialAccessCost   = *FocalProximity of BigRoad spreading in InitialTransportCost at ...*

Fig. 8-6 presents a layer indicating the amount of timber that would ultimately be transported over each location if each of the harvestable locations were to be accessed by way of its own minimum-cost path. This layer was created from the Brown's *Vegetation* layer and *InitialAccessCost* as follows:

Timber   = *LocalRating of Vegetation with 1 for 2 3*
WhichWay   = *IncrementalDrainage of InitialAccessCost*
InitialTraffic   = *FocalSum of Timber spreading through WhichWay at ...*
InitialTraffic   = *LocalRating of BigRoad with -0 for 1 with InitialTraffic for -0*

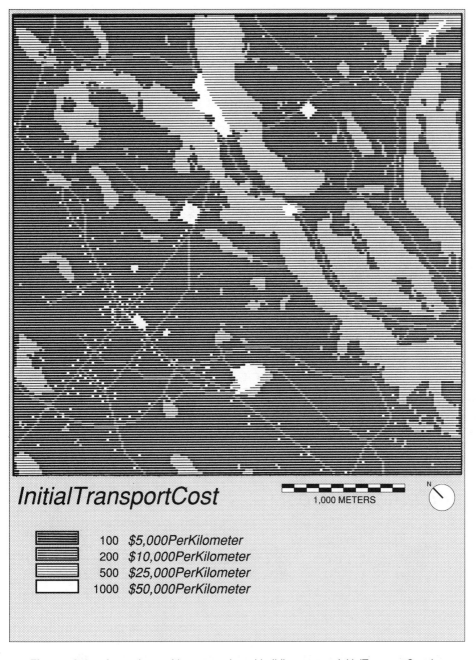

InitialTransportCost

1,000 METERS

N

	100	*$5,000PerKilometer*
	200	*$10,000PerKilometer*
	500	*$25,000PerKilometer*
	1000	*$50,000PerKilometer*

Figure 8-4 A map layer of incremental road-building costs. *InitialTransportCost* is a layer indicating the hypothetical dollar cost of constructing a logging road across each location in the Brown's Pond study area.

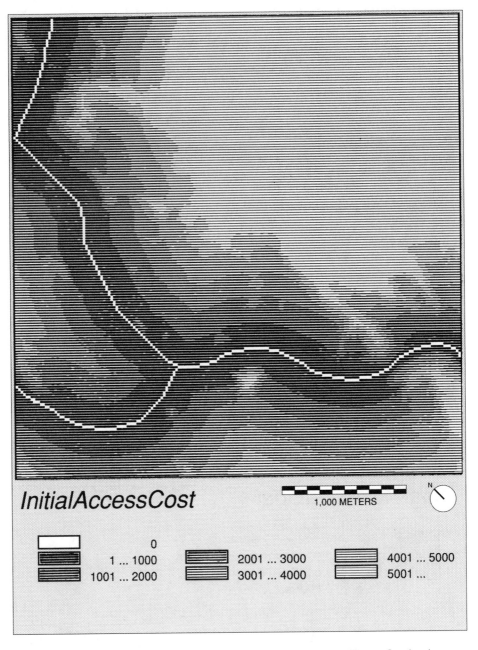

Figure 8-5 A map layer of cumulative road-building costs. *InitialAccessCost* is a layer indicating the minimum travel-cost distance in dollars from each location in the Brown's Pond study area to a major road. Note that each shading pattern represents not just one zone but a range of costs.

InitialTraffic

1,000 METERS

N

▬ -0		
☐ 0	▤ 11 ... 20	▤ 31 ... 40
▤ 1 ... 10	▤ 21 ... 30	▤ 41 ...

Figure 8-6   A map layer of estimated timber-hauling traffic. *InitialTraffic* is a layer indicating the amount of timber (number of harvested locations) that would pass through each location within the Brown's Pond study area if every wooded location were to be accessed by way of that path that minimizes its own access cost. Note that each shading pattern represents not just one zone but a range of values.

One problem with this should be apparent from Fig. 8-6. While the road layout suggested by *InitialTraffic* is indeed optimal from the individual perspective of any one harvestable location, it is not at all optimal from the broader perspective of the entire study area. This is because there has been no attempt to access multiple locations with shared stretches of roadway.

To develop this kind of road network, however, we must again contend with the kind of chicken-and-egg relationship that is characteristic of holistic allocation problems. The dilemma in this case arises from the fact that efficient routes cannot be established without at least some indication of the degree to which portions of those routes will serve multiple locations. This, of course, is something that cannot be known with certainty until all routes are in place.

The key to a solution for this holistic dilemma once again lies not in a quest for certainty but in the use of heuristic assumptions. We can begin by tentatively accepting the information presented in Fig. 8-6 as a rough approximation of the general pattern of timber traffic that is likely to flow from the study area when a road network is finally established. We can then use this information to distribute construction costs among locations likely to be served. To do so, we need only equate higher values on the *InitialTraffic* layer with lower values on a revised version of *InitialTransportCost* layer. If this revised estimate of incremental road-building costs is stored as a layer entitled *SharedTransportCost*, then an operation given as

SharedAccessCost = *FocalProximity of BigRoad spreading in SharedTransportCost at ...*

can be used to generate a revised version of the *InitialAccessCost* layer shown in Fig. 8-5. This new version, shown in Fig. 8-7, differs from the original in that travel costs accumulate more gradually toward areas that are likely to generate more traffic. It is these areas that are most likely to present opportunities for well-shared roads.

To begin to see the emerging pattern of this shared-road network, we can now revise the *InitialTraffic* layer shown in Fig. 8-6. By using the *SharedAccessCost* layer shown in Fig. 8-7, we can generate the *SharedTraffic* layer shown in Fig. 8-8. Note that the dendritic pattern of *SharedTraffic* is coarser than that of *InitialTraffic*. Traffic volume has shifted from a larger number of low-volume branches to a smaller number of high-volume trunk lines. If we repeat these steps such that each new access cost layer is computed from the previous traffic layer, we will eventually arrive at a traffic layer like that shown in Fig. 8-9. To transform this into a road network, we need only decide on the level of traffic that warrants road construction.

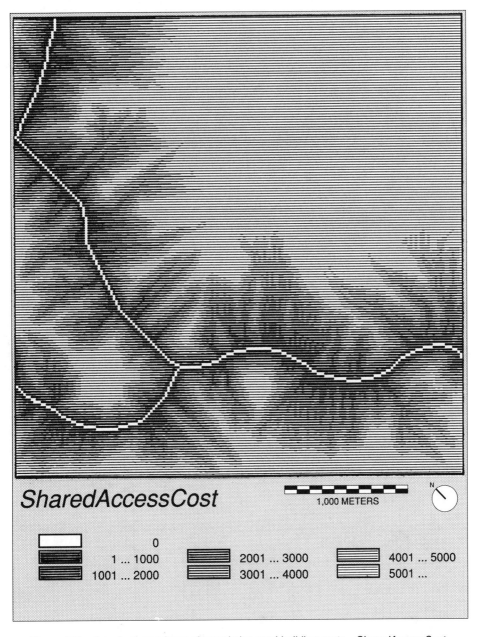

Figure 8-7   A revised map layer of cumulative road-building costs. *SharedAccessCost* is a layer indicating the travel-cost distance in dollars from each location in the Brown's Pond study area to a major road. These distances reflect an attempt to minimize the overall sum of such costs. Note that each shading pattern represents not just one zone but a range of costs.

Figure 8-8 A revised map layer of estimated timber-hauling traffic. *SharedTraffic* is a layer indicating the amount of timber (number of harvested locations) that would pass through each location within the Brown's Pond study area if all wooded locations were to be accessed by paths that begin to minimize shared access costs. Note that each shading pattern represents not just one zone but a range of values.

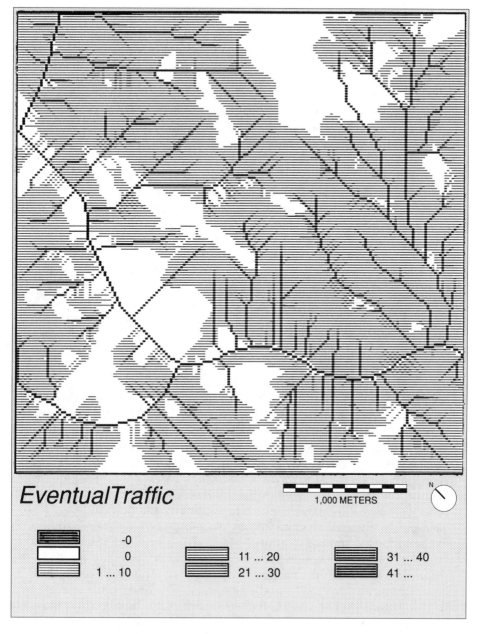

EventualTraffic

1,000 METERS

N

Legend:
- -0
- 0
- 1 ... 10
- 11 ... 20
- 21 ... 30
- 31 ... 40
- 41 ...

Figure 8-9 Another revised map layer of estimated timber-hauling traffic. *EventualTraffic* is a layer indicating the amount of timber (number of harvested locations) that would pass through each location within the Brown's Pond study area if all wooded locations were to be accessed by paths that minimize shared access costs. Note that each shading pattern represents not just one zone but a range of values.

We have so far considered a series of holistic allocation problems that have involved issues ranging from size, sparsity, and density to conterminous form, minimum-cost paths, and efficient access networks. Note that the holistic qualities involved in these problems have gradually become more specific. Note, too, that the allocation processes required to achieve these qualities have gradually become more complex. As holistic allocation problems become even more specific in terms of the qualities they are intended to achieve, techniques for the solution of these problems rapidly become quite a bit more sophisticated (and well beyond the scope of this text). Such problems can often be at least partially solved, however, by adopting the general approach to holistic allocation so far described.

This approach can also be effective in generating solutions to problems whose holistic characteristics go beyond those that are associated with the allocation of a single zone in response to a single criterion. We have already seen, for example, how multiple criteria that cannot be combined atomistically can nonetheless often be reconciled by making a series of tentative allocations, each responding to the holistic qualities that arise from the one before. The same technique can also be used to allocate multiple zones when the siting criteria for two or more zones are holistically interrelated. This will be the case whenever allocation criteria are such that neither of two zones can be sited definitively until the other is sited first.

Suppose, for example, that a housing development and a sewage treatment plant are to be allocated as part of a multiple land use plan. Such a plan might also include roads, commercial and industrial sites, agricultural land, recreation facilities, conservation areas, and so on. In siting the housing as part of this plan, factors to be considered might well include existing site characteristics such as soil type, population density, or distance to the nearest road. They might also include site characteristics that do not yet exist but do relate to other yet-to-be-sited land uses. Site suitability for new housing, for example, might well be enhanced by close-but-not-too-close proximity to the proposed sewage treatment plant. To incorporate this criterion into the housing allocation process, we would have to site the treatment plant first and then measure distance to it. The result could then be treated as an "existing" site condition.

Before we can allocate that treatment plant, however, we must consider its own allocation criteria. Just like those used to allocate housing, these criteria will probably relate not only to existing site conditions but also to other land uses that have yet to be allocated. They might well state, for example, that the treatment plant should be situated near (or downhill of, or within the same administrative region as, or in some other relationship to) the new housing development. If this is the case (or if the treatment plant relates to any other proposed land use(s) affected by housing), then that housing must be allocated before the plant can be sited for the same reason that the plant must be sited before allocating housing. It is one more chicken-and-egg situation like those we have seen in other holistic allocation problems, and it is amenable to the same kind of solution.

We begin by separating the atomistic and holistic criteria associated with each land use. For each of these uses, we initially generate a map layer on which every location is characterized in terms of the use's atomistic allocation criteria. These layers provide the basis for a first round of allocations.

In this first round, we allocate each land use independently and make no attempt to relate any one land use to any other. Two or more uses could well be allocated to the same location. While these initial allocations are therefore not likely to be workable, they do at least provide an indication of where each land use is inclined to settle in response to some of its siting criteria.

They also provide the basis for a second round of allocations, a round that can now begin to address holistic interrelationships. To do this, we must suspend any skepticism for a moment and assume that each land use will actually occur where initially allocated. In this way, we can at least tentatively treat those land uses as "existing" site conditions and measure whatever holistic spatial characteristics they may exhibit. We can then express these measurements as atomistic site characteristics and combine them with the original atomistic siting criteria to generate a revised set of site suitability layers.

In the case of that housing development, for example, its relationship to the new sewage treatment plant would be addressed by mapping all of those areas that are close-but-not-too-close to any of the potential treatment plant sites initially allocated. Once mapped, this factor would be combined with the atomistic factors initially involved to generate an updated layer of site suitability for housing.

When this has been done for all of the proposed land uses, a second round of allocations can proceed much like the first. Each land use is again assigned to whatever locations are deemed most suitable. This round will differ from the first, however, in that the siting criteria for each land use can now respond to the tentative siting of every other land use. As a result, the number of conflicts between allocated uses will tend to be reduced, and locations that satisfy holistic as well as atomistic siting criteria will tend to be favored.

This update-and-reallocate process is then repeated again and again, each time using the land use locations selected in the previous round to revise the site suitability layers upon which new allocations will be based. Though there is no guarantee that a land use at a particular location in one round will still be there in the next, this does tend to become more and more likely each time the process is repeated. And eventually a more or less stable solution will be reached. This solution will seldom be demonstrably optimal but will almost certainly be reasonable.

The iterative nature of this type of allocation is also significant in another way. It provides for a gradual progression from siting issues of a general nature to those that are much more specific. And it does so in a way that can easily be modified at any point. This ability to intervene in the allocation process makes it possible to incorporate the kind of human judgment that can never be fully automated.

Evaluating Results

Once a zone or a set of zones has been allocated in response to holistic criteria, it remains to evaluate the degree to which those criteria have been satisfied. The process is similar to that involved in atomistic evaluation. We must estimate the levels of environmental quality that result from allocated zones by subjecting those zones (and only those zones) to the descriptive modeling procedures that were used to establish the siting criteria. In the case of atomistic allocation, this could be done by simply retrieving a previously calculated quality estimate for each of the locations to which a given zone had been allocated. In the case of holistic allocation, however, the process is more involved. Here, we cannot rely on previous calculations but must reapply quality-measuring procedures to the allocated zones.

To illustrate this, consider the case in which the siting criterion for a zone to be allocated states that it must be more or less circular. Without a specific set of locations to be evaluated as a whole, this holistic criterion so easily stated can never be applied. Once a candidate set of locations is available for consideration, however, it is relatively easy to determine whether or not they form a circle.

But how should the measure of this holistic quality then be expressed? Should a single measure of the allocated zone's overall circularity (perhaps the ratio of the square root of its area to its perimeter) be assigned to each of its constituent locations? Should each location be characterized in terms of the degree to which it contributes to or detracts from the overall form of a circle? Or should this sort of assessment be applied to locations outside of the allocated zone as well as within it?

While the answer to this type of question will certainly vary from one situation to another, it is generally best to design evaluations such that they do more than merely indicate how well or how poorly a criterion has been satisfied. Ideally, an evaluation should also provide the basis for reallocation.

Note that this is not an issue in atomistic problems where criteria are defined, zones are allocated, and allocations are evaluated on a location-by-location basis. This localization of cause and effect makes it easy to move from "what's good" or "what's bad" to "what should be done about it."

In the case of holistic allocation, on the other hand, it is much more difficult to achieve this kind of direct relationship between cause and effect. It is nonetheless often possible to express at least certain holistic qualities in terms of atomistic increments. The timber harvesting road allocation technique is a good example.

The timber harvesting example embodies what may, in fact, be the two most important fundamentals of cartographic modeling technique. The first of these is adoption of an atomistic perspective. The second is a willingness to rely on heuristic methods. By expressing all manner of geographic phenomena in terms of the same elementary units, the diverse qualities of those phenomena can be fully and readily interrelated. And by accepting the risk of uncertainty, we can move from problems to solutions in a way that reflects and even engages those creative powers that remain so distinctly human. Though cartographic modeling alone can do little to satisfy the kind of concern that is reflected in T. S. Eliot's quest for *wisdom lost in knowledge* and *knowledge lost in information*, it can do much to enlighten the more basic quest for *information lost in data* and *data lost in experience*.

8-3 QUESTIONS

This chapter has described a range of techniques by which descriptive models can be inverted into prescriptive form. The following questions explore the application of these techniques.

8-1 The dendritic pattern of the *EventualTraffic* layer presented in Fig. 8-9 is similar to those of trees, blood vessels, and other natural phenomena involving growth. Why?

8-2 Given a layer of the roads and homes involved, how would you allocate a newspaper delivery route?

8-3 Given a layer indicating the locations of trees on a forest property, how would you route a piece of heavy equipment requiring a berth of at least two meters from one location to another without disturbing any of these trees?

8-4 How would the routing technique illustrated in Fig. 7-3 have to be modified to allocate a power transmission line, given that the major siting costs are those associated with towers that may be separated by as much as 50 meters?

8-5 How would you attempt to allocate the path between two specified locations that minimizes topographic slope? How about a path that not only minimizes incremental costs but is also of a specified length? And how about one that, instead of minimizing incremental costs, maximizes the overall variety of conditions through which it passes? What makes these seemingly minor variations on the path-allocating technique illustrated in Fig. 7-3 so different (and so difficult)?

8-6 The drill sergeant glares squarely in the face of the new recruit and barks, "Line up, soldier!" How is this individual (the smart-aleck recent graduate of a program in geographic information systems and cartographic modeling) to interpret and respond to such a problem in holistic spatial allocation?

8-7 Given a layer depicting each location's atomistic suitability for inclusion within a park system and given a layer of existing park land, how would you allocate new park land in response to a criterion stating that the overall park system should ultimately be in one piece? How about one piece of exactly 200 hectares?

8-8 Given map layers of sensitive viewers, scenic objects, and visual obstructions, how could you identify those obstructions that would result in scenic views if removed?

8-9 Each of five map layers depicts an alternative plan of some ten proposed land uses for a given study area. These plans are all based on the same set of ten site suitability layers, one for each of the proposed land uses. How should these alternatives be compared? Suppose that the siting criteria for each proposed land use were expressed not only in the form of a site suitability layer, but also in terms of a procedure describing its ideal relationship to the other proposed land uses?

8-10 How does the multiple land use allocation strategy described at the end of this chapter relate to the way in which seating patterns in a classroom tend to emerge during the first few days of a school year?

8-11 One way to anticipate conflicts in the multiple land use allocation process described in this chapter is to start each iteration by generating a "competition" layer. This would be a layer on which each location's value indicates the number of uses for which that location has been deemed highly suitable. How could this information then be used to mitigate allocation conflicts? How could it be used instead to "maintain options" in the allocation process or to "strengthen negotiating position"?

8-12 Just what are the significant distinctions among *experience, data, information, knowledge,* and *wisdom* in a geographic context?

SELECTED READINGS

Cartographic modeling is, a field in which narrow and rapidly growing shoots have begun to spring from broad and well-established roots. This makes concise, coherent, and comprehensive bibliography a challenge. The following selections represent only a first response to that challenge. They comprise what amounts to an entree to the literature on this field's precursors, principles, practices, and prospects.

Aronoff, Stanley, *Geographic Information Systems: A Management Perspective*, Ottawa: WDL Publications, 1989. A substantial introduction to practical aspects of geographic information systems.

Burrough, Peter A., *Principles of Geographical Information Systems for Land Resources Assessment*, Oxford: Clarendon Press, 1986. An introduction to the broader field of which cartographic modeling is a part.

Chan, Kelly, *Evaluating Descriptive Spatial Models for Prescriptive Inference*, Unpublished doctoral dissertation, Harvard University, 1988. A formalization of cartographic modeling concepts in terms of entity relations rather than mathematical functions, with particular emphasis on implications for prescriptive modeling.

Davis, John C., *Statistics and Data Analysis in Geology*, New York: John Wiley and Sons, Inc., 1973. A clear introduction to some of the more advanced techniques in descriptive cartographic modeling from the perspective of geological applications.

McHarg, Ian L., *Design with Nature*, Garden City, N.Y.: Natural History Press, 1969. The classic introduction to overlay mapping.

GIS AND CARTOGRAPHIC MODELING

Monmonier, Mark S., *Computer-assisted Cartography: Principles and Prospects*, Englewood Cliffs, N.J.: Prentice-Hall, 1982. Introduction to a variety of techniques in digital mapping from a cartographic perspective.

Ripple, William J., ed., *Geographic Information Systems for Resource Management: A Compendium*, Falls Church, Va.: American Society for Photogrammetry and Remote Sensing and American Congress on Surveying and Mapping, 1987. A compilation of reports on geographic information system applications oriented toward descriptive modeling.

Rosenfeld, Azriel, and Avinash C. Kak, *Digital Picture Processing*, New York: Academic Press, 1976. The prominent text in a field whose methods (if not its objectives) are closely related to those of cartographic modeling.

Star, Jeffrey and John E. Estes, *Geographic Information Systems: An Introduction*, Englewood Cliffs, N.J.: Prentice-Hall, Inc., 1990. An introductory review of general issues in the field of geographic information systems with particular emphasis on their relationship to remote sensing techniques.

Teicholz, Eric, and Brian J. L. Berry, eds., *Computer Graphics and Environmental Planning*, Englewood Cliffs, N.J.: Prentice-Hall Inc., 1983. A collection of case studies oriented toward prescriptive applications of geographic information systems.

Tomlin, C. Dana, *Digital Cartographic Modeling Techniques in Environmental Planning*, Unpublished doctoral dissertation, Yale University, 1983. The immediate predecessor to this text.

Tomlin, C. Dana, *The IBM Personal Computer Version of the Map Analysis Package*, Cambridge: Harvard University Laboratory for Computer Graphics and Spatial Analysis, 1985. Manual for the software used to generate the examples presented throughout this text.

Tomlin, Sandra M., *Timber Harvest Scheduling and Spatial Allocation*, Unpublished master's thesis, Harvard University, 1981. Source of the timber-harvesting road allocation technique presented in Chapter 8.

Unwin, David, *Introductory Spatial Analysis*, London: Methuen and Co. Ltd., 1981. A primer on statistical methods quite complementary to those of cartographic modeling.

Appendix

CARTOGRAPHIC MODELING
OPERATIONS

The fundamental capabilities of a cartographic modeling language can be expressed in terms of its operations for data interpretation. The following pages summarize these capabilities by describing the form and the function of statements for operations

- *FocalFUNCTION*,
- *IncrementalFUNCTION*,
- *LocalFUNCTION*, and
- *ZonalFUNCTION*.

This material is not intended to introduce or to fully explain each operation but merely to serve as a reference.

A-1   *FocalFUNCTION* OPERATIONS

The *FocalFUNCTION* operations are introduced in Chapter 5. Each generates a new map layer on which every location is set to a value computed as a specified function of the values, distances, and/or directions of neighboring locations. Neighborhood distances may be measured in terms of physical separation, travel costs, or lines of sight. *FocalFUNCTION* statements are specified as follows, where *spreading* and *radiating* phrases are mutually exclusive:

```
NEWLAYER          =  FocalFUNCTION of FIRSTLAYER
                     [at DISTANCE] etc.   [by DIRECTION] etc.
                     [spreading [in FRICTIONLAYER]
                          [on SURFACELAYER]
                            [through NETWORKLAYER] ]
                     [radiating [on SURFACELAYER]
                          [from TRANSMISSIONLAYER]
                          [through OBSTRUCTIONLAYER]
                          [to RECEPTIONLAYER] ]
```

The NEWLAYER Phrase

NEWLAYER is the title to be assigned to the new map layer. If NEWLAYER matches the title of an existing map layer, the existing layer is deleted when the new layer is created. The resolution and orientation of NEWLAYER are those of FIRSTLAYER.

The = FocalFUNCTION Phrase

= FocalFUNCTION specifies the function to be used in calculating new values. FUNCTION is one of the terms Bearing, Combination, Gravitation, Insularity, Majority, Maximum, Mean, Minimum, Minority, Neighbor, Percentage, Percentile, Product, Proximity, Ranking, Rating, Sum, or Variety.

= FocalBearing specifies that each location's NEWLAYER value should indicate the direction of the nearest location within its neighborhood whose FIRSTLAYER value is other than -0. Directions are measured as shown in Fig. 1-15 and are expressed in clockwise degrees from north (as indicated by FIRSTLAYER's orientation) such that values of 90, 180, 270, and 360 indicate bearings of due east, south, west, and north, respectively. If two or more nearest neighbors are found, the highest of their bearing values is assigned. If no neighbor is found, -0 is assigned. The spreading option has no effect on this function.

= FocalCombination specifies that each location's NEWLAYER value should indicate the particular combination of FIRSTLAYER zones that occur at one or more locations within its neighborhood. Values of one, two, three, and so on are assigned according to the order in which neighborhoods occur when sorted by minimum value.

= FocalGravitation specifies that each location's NEWLAYER value should indicate the inverse-square-distance-weight average of the FIRSTLAYER values at all locations within its neighborhood.

= *FocalInsularity* specifies that each location's NEWLAYER value should uniquely match the NEWLAYER value assigned to all other locations within its neighborhood that are also within the same FIRSTLAYER zone. Values of one, two, three, and so on are assigned to spatially continuous or near-continuous groups of locations within each FIRSTLAYER zone according to the order in which they occur when sorted by row-then-column coordinates.

= *FocalMajority* specifies that each location's NEWLAYER value should indicate which FIRSTLAYER value occurs most often within its neighborhood. If two or more such values are found, -0 is assigned.

= *FocalMaximum* specifies that each location's NEWLAYER value should indicate the highest FIRSTLAYER value within its neighborhood.

= *FocalMean* specifies that each location's NEWLAYER value should indicate the average of the FIRSTLAYER values at all locations within its neighborhood.

= *FocalMinimum* specifies that each location's NEWLAYER value should indicate the lowest FIRSTLAYER value within its neighborhood.

= *FocalMinority* specifies that each location's NEWLAYER value should indicate which FIRSTLAYER value occurs least often within its neighborhood. If two or more such values are found, -0 is assigned.

= *FocalNeighbor* specifies that each location's NEWLAYER value should indicate the FIRSTLAYER value of the nearest location within its neighborhood whose value is other than -0. If no neighbor is found, -0 is assigned.

= *FocalPercentage* specifies that each location's NEWLAYER value should indicate the number of locations within its neighborhood that have equal FIRSTLAYER values. This number is expressed as a percentage of the total number of locations in that neighborhood.

= *FocalPercentile* specifies that each location's NEWLAYER value should indicate the number of locations within its neighborhood that have lower FIRSTLAYER values. This number is expressed as a percentage of the total number of locations in that neighborhood.

= *FocalProduct* specifies that each location's NEWLAYER value should indicate the multiplicative product of the FIRSTLAYER values at all locations within its neighborhood.

= *FocalProximity* specifies that each location's NEWLAYER value should indicate the distance of the nearest location within its neighborhood whose value is other than -0. This value, normally set to zero, is treated as a measure of distance "already consumed" at that location as new distances are calculated. If no neighbor is found, -0 is assigned.

= *FocalRanking* specifies that each location's NEWLAYER value should indicate the number of zones occurring at one or more locations within its neighborhood that have lower FIRSTLAYER values.

= *FocalRating* specifies that each location's *NEWLAYER* value should be explicitly assigned according to the combination of *FIRSTLAYER* zones occurring within its neighborhood. New values are assigned through one or more phrases given as *with NEWVALUE for FIRST-ZONES* where

- *NEWVALUE* is a numeral or the title of an existing layer, and
- *FIRSTZONES* is one or more numerals representing *FIRSTLAYER* values.

As a numeral, *NEWVALUE* specifies a constant value to be applied to the focus of any neighborhood that contains (only) the specified *FIRSTLAYER* value(s). As a title, *NEWVALUE* indicates that each location's new value is to be set to that of the same location on the map layer whose title is specified. If no new value is specified for a particular neighborhood, its focus is set to -0.

= *FocalSum* specifies that each location's *NEWLAYER* value should indicate the sum of the *FIRSTLAYER* values at all locations within its neighborhood.

= *FocalVariety* specifies that each location's *NEWLAYER* value should indicate the number of *FIRSTLAYER* zones that occur at one or more locations within its neighborhood.

The *of FIRSTLAYER* Phrase

of FIRSTLAYER specifies the existing map layer whose values are to be summarized within each neighborhood. *FIRSTLAYER* is this layer's title or a numeral. As a numeral, it represents a layer on which all locations are set to that numeral's value.

The *at DISTANCE* Phrase

at DISTANCE specifies the distance(s) at which locations in the vicinity of a given location are to be treated as part of its neighborhood. *DISTANCE* is one or more numerals and/or titles. Numerals represents constant distances. Titles represent distances that may vary from one neighborhood focus to another according to the values of the existing layer whose title is specified. In either case, distances are expressed in the units of *FIRSTLAYER*'s resolution unless a *FRICTIONLAYER* is specified. If no *DISTANCE* is specified, each location's neighborhood extends only to its adjacent locations.

The *by DIRECTION* Phrase

by DIRECTION specifies the direction(s) at which those locations in the vicinity of a given location are to be treated as part of its neighborhood. *DIRECTION* is one or more numerals and/or titles. Numerals represent constant directions. Titles represent directions that may vary from one neighborhood focus to another according to the values of the existing layer whose title is specified. In either case, directions are measured from the neighborhood focus and are expressed in clockwise degrees from north (as indicated by *FIRSTLAYER*'s orientation) such that 90 (or -270 or 450), 180, 270, and 360 represent directions facing due east, south, west, and north, respectively. When *by DIRECTION* immediately follows an *at DISTANCE* phrase, it refers only to those locations that satisfy those particular *at DISTANCE* conditions. Otherwise, it refers to all locations. If no *DIRECTION* is specified, each location's neighborhood extends in all directions.

The *spreading* Phrase

spreading specifies that all distances from each neighborhood focus are to be measured as an accumulation of costs associated with location-to-location movement. *spreading* may be specified alone or with up to three modifying phrases including
- *in FRICTIONLAYER*,
- *on SURFACELAYER*, and
- *through NETWORKLAYER*.

spreading in FRICTIONLAYER specifies the title of an existing layer whose values indicate the "width" of each location, not in feet or meters, but in units such as minutes or dollars that can accumulate at rates that vary from one location to another. These units are then used to measure neighborhood distances such that the distance from any one location to an adjacent neighbor would be the average of their *FRICTION-LAYER* values (multiplied by 1.4142 if that neighbor were diagonally adjacent).

spreading on SURFACELAYER specifies the title of an existing map layer on which each location's value defines its vertical position (in units corresponding to those of *FIRSTLAYER*'s resolution) on a surface. This surface is then used to measure neighborhood distances such that the distance from any one location to an adjacent neighbor would be increased by the secant of the vertical angle between them.

spreading through NETWORKLAYER specifies the title of an existing map layer on which each location's value identifies those adjacent neighbors to which it is "connected." These connections are then used to measure neighborhood distances such that the distance from any one location to an adjacent neighbor is infinite unless they are connected. Each *NETWORKLAYER* value is encoded as a sum of 1, 2, 4, 8, 16, 32, 64, and/or 128 representing upper left, upper, upper right, left, right, lower left, lower, and upper right neighbors, respectively.

The *radiating* Phrase

radiating specifies that distances from each neighborhood focus are to be measured over unobstructed lines of sight. *radiating* may be specified alone or with up to four modifying phrases including

- *on SURFACELAYER*,
- *from TRANSMISSIONLAYER*,
- *through OBSTRUCTIONLAYER*, and
- *to RECEPTIONLAYER*.

radiating on SURFACELAYER specifies an existing map layer on which each location's value defines its vertical position (in units corresponding to those of *FIRSTLAYER*'s resolution) on an opaque surface. This is normally a surface of topographic elevations. If no *SURFACELAYER* is specified, a horizontal surface is assumed.

radiating from TRANSMISSIONLAYER specifies an existing map layer on which each location's value indicates the height at which lines of sight are to be "transmitted." These heights are measured in units corresponding to those of *FIRSTLAYER*'s resolution, and each location's height is measured above its *SURFACELAYER* position. If no *TRANSMISSIONLAYER* is specified, all transmitting heights are set to zero.

radiating through OBSTRUCTIONLAYER specifies the title of an existing map layer on which each location's value indicates the height below which no line of sight can pass over that location's grid square. These heights are measured in units corresponding to those of *FIRSTLAYER*'s resolution, and each location's height is measured above its *SURFACELAYER* position. If no *OBSTRUCTIONLAYER* is specified, all obstructing heights are set to zero.

radiating to RECEPTIONLAYER specifies the title of an existing map layer on which each location's value indicates the height at (or below) which a line of sight can be received. These heights are measured in units corresponding to those of *FIRSTLAYER*'s resolution, and each location's height is measured above its *SURFACELAYER* position. If no *RECEPTIONLAYER* is specified, all receiving heights are set to zero.

A-2 *IncrementalFUNCTION* **OPERATIONS**

The *IncrementalFUNCTION* operations are introduced in Chapter 5. Each generates a new map layer on which every location is set to a value characterizing whatever portion of a lineal, areal, or surficial condition may be represented by that location. *IncrementalFUNCTION* statements are specified as

NEWLAYER   = *IncrementalFUNCTION [of FIRSTLAYER] [on SURFACELAYER]*

The *NEWLAYER* **Phrase**

NEWLAYER is the title to be assigned to the new map layer. If *NEWLAYER* matches the title of an existing map layer, the existing layer is deleted when the new layer is created. The resolution and orientation of *NEWLAYER* are those of *FIRSTLAYER*.

The *IncrementalFUNCTION* **Phrase**

IncrementalFUNCTION specifies the function to be used in calculating new values. *FUNCTION* is one of the terms *Area, Aspect, Drainage, Frontage, Gradient, Length, Linkage, Partition,* or *Volume.*

= *IncrementalArea* specifies that each location's *NEWLAYER* value should indicate the area of whatever portion of an areal condition is represented by that location on *FIRSTLAYER* when projected onto a surface inferred from *SURFACELAYER* values. Area is in the (squared) units of *FIRSTLAYER*'s resolution.

= *IncrementalAspect* specifies that each location's *NEWLAYER* value should indicate the compass direction of steepest descent for a plane inferred from the *SURFACELAYER* values of that location and any adjacent neighbors that share its *FIRSTLAYER* value. Aspect is expressed in clockwise degrees from north (as indicated by *SURFACELAYER*'s orientation) such that values of 90, 180, 270, and 360 represent slopes facing east, south, west, and north, respectively, while zero represents a horizontal surface.

= *IncrementalDrainage* specifies that each location's *NEWLAYER* value should indicate which of its adjacent neighbors lie(s) upstream on a surface inferred from the *SURFACELAYER* values of all locations in the same *FIRSTLAYER* zone. The upstream neighbors of a given location are those from which the location is in a direction of steepest descent. Specific combinations of neighbors are encoded as shown in Fig. 5-12.

= *IncrementalFrontage* specifies that each location's NEWLAYER value should indicate the length of the boundary(ies) formed by whatever portion of an areal condition is represented by that location on FIRSTLAYER when projected onto a surface inferred from SURFACELAYER values. Lengths are in the units of FIRSTLAYER's resolution.

= *IncrementalGradient* specifies that each location's NEWLAYER value should indicate the slope of a plane inferred from the SURFACELAYER values of that location and whatever adjacent neighbors share its FIRSTLAYER value. Slope is expressed in degrees.

= *IncrementalLength* specifies that each location's NEWLAYER value should indicate the length of whatever portion of a lineal condition is represented by that location on FIRSTLAYER when projected onto a surface inferred from SURFACELAYER values. Lengths are in the units of FIRSTLAYER's resolution.

= *IncrementalLinkage* specifies that each location's NEWLAYER value should indicate the type of form inferred from the FIRSTLAYER values of that location and its adjacent neighbors.

= *IncrementalPartition* specifies that each location's NEWLAYER value should indicate the type of areal boundary inferred from its FIRSTLAYER value and those of its upper, upper right, and right neighbors.

= *IncrementalVolume* specifies that each location's NEWLAYER value should indicate the surficial volume beneath whatever portion of an areal condition is represented by that location's FIRSTLAYER value when projected onto a surface inferred from SURFACELAYER values. Surficial volume is in the (cubed) units of SURFACELAYER's resolution.

The *of FIRSTLAYER* Phrase

of FIRSTLAYER specifies the existing map layer whose values define the form of lineal or areal conditions as shown, respectively, in Fig. 1-17 and 1-19. FIRSTLAYER is this layer's title. If no FIRSTLAYER is specified, form is inferred from a layer on which all locations are in the same zone. In this case, an *on SURFACELAYER* phrase must be specified.

The *on SURFACELAYER* Phrase

on SURFACELAYER specifies the existing map layer whose values define the vertical position of each location on a surface as shown in Fig. 1-21. SURFACELAYER is this layer's title. If no SURFACELAYER is specified, all locations are assumed to be at the same vertical position. In this case, an *of FIRSTLAYER* phrase must be specified.

A-3 *LocalFUNCTION* OPERATIONS

The *LocalFUNCTION* operations are introduced in Chapter 4. Each generates a new map layer on which every location is set to a value computed as a specified function of the location's value(s) on one or more existing map layers. *LocalFUNCTION* statements are specified as

NEWLAYER = *LocalFUNCTION of FIRSTLAYER [and NEXTLAYER] etc.*

The *NEWLAYER* Phrase

NEWLAYER is the title to be assigned to the new map layer. If *NEWLAYER* matches the title of an existing map layer, the existing layer is deleted when the new layer is created. The resolution and orientation of *NEWLAYER* are those of *FIRSTLAYER*.

The = *LocalFUNCTION* Phrase

= *LocalFUNCTION* specifies the function to be used in calculating new values. *FUNCTION* is one of the terms *ArcCosine, ArcSine, ArcTangent, Combination, Cosine, Difference, Majority, Maximum, Mean, Minimum, Minority, Product, Rating, Ratio, Root, Sine, Sum, Tangent,* or *Variety* .

= *LocalArcCosine* specifies that each location's *NEWLAYER* value should indicate the angle whose cosine is one hundredth of that location's *FIRSTLAYER* value. This angle is expressed in degrees ranging from zero to 180.

= *LocalArcSine* specifies that each location's *NEWLAYER* value should indicate the angle whose sine is one hundredth of that location's *FIRSTLAYER* value. This angle is expressed in degrees ranging from -90 to 90.

= *LocalArcTangent* specifies that each location's *NEWLAYER* value should indicate the angle whose tangent is one hundredth of that location's *FIRSTLAYER* value divided by its *NEXTLAYER* value(s), if any. This angle is expressed in degrees ranging from -90 to 90.

= *LocalCombination* specifies that each location's *NEWLAYER* value should indicate the particular combination of zones associated with that location on *FIRSTLAYER* and any *NEXTLAYER*(s) specified. Values of one, two, three, and so on are assigned according to the order in which combinations occur when sorted by minimum value.

= *LocalCosine* specifies that each location's *NEWLAYER* value should indicate the cosine (multiplied by 100) of the angle represented in degrees by the location's *FIRSTLAYER* value.

= *LocalDifference* specifies that each location's *NEWLAYER* value should be computed by subtracting its *NEXTLAYER* value(s) from its *FIRSTLAYER* value.

= *LocalMajority* specifies that each location's *NEWLAYER* value should indicate which of its *FIRSTLAYER* and *NEXTLAYER* values occurs most often. If two or more such values are found, -0 is assigned.

= *LocalMaximum* specifies that each location's *NEWLAYER* value should indicate the highest of its *FIRSTLAYER* and *NEXTLAYER* values.

= *LocalMean* specifies that each location's *NEWLAYER* value should indicate the average of its *FIRSTLAYER* and *NEXTLAYER* values.

= *LocalMinimum* specifies that each location's *NEWLAYER* value should indicate the lowest of its *FIRSTLAYER* and *NEXTLAYER* values.

= *LocalMinority* specifies that each location's *NEWLAYER* value should indicate which of its *FIRSTLAYER* and *NEXTLAYER* values occurs least often. If two or more such values are found, -0 is assigned.

= *LocalProduct* specifies that each location's *NEWLAYER* value should indicate the multiplicative product of its *FIRSTLAYER* and *NEXTLAYER* values.

= *LocalRating* specifies that each location's *NEWLAYER* value should be explicitly assigned according to the *FIRSTLAYER* and any *NEXTLAYER* zone(s) containing that location. New values are assigned through one or more phrases given as *with NEWVALUE for FIRSTZONES [on NEXTZONES]* where

- *NEWVALUE* is a numeral or the title of an existing layer,
- *FIRSTZONES* is one or more numerals representing *FIRSTLAYER* values, and
- *NEXTZONES* is one or more numerals representing *NEXTLAYER* values.

After each *for FIRSTZONES* phrase, a separate *on NEXTZONES* phrase must be given for each *NEXTLAYER* specified. As a numeral, *NEWVALUE* specifies a constant value to be applied to all locations exhibiting the *FIRSTLAYER* and *NEXTLAYER* values (if any) specified. As a title, *NEW-VALUE* indicates that each location's new value is to be set to that of the same location on the map layer whose title is specified. If no new value is specified for the zone(s) containing a particular location, that location retains its *FIRSTLAYER* value unless a *NEXTLAYER* was also specified, in which case -0 is assigned.

= *LocalRatio* specifies that each location's *NEWLAYER* value should be computed by dividing its *FIRSTLAYER* value by its *NEXTLAYER* value(s). This is done such that division of zero by zero yields zero,while division of any other positive value by zero yields ++, and division of any negative value by zero yields --.

= *LocalRoot* specifies that each location's *NEWLAYER* value should be set to whatever root of its *FIRSTLAYER* value is indicated by its *NEXTLAYER* value (or the product of its *NEXTLAYER* values, if more than one).

= *LocalSine* specifies that each location's *NEWLAYER* value should indicate the sine (multiplied by 100) of the angle represented in degrees by the location's *FIRSTLAYER* value.

= *LocalSum* specifies that each location's *NEWLAYER* value should indicate the sum of its *FIRSTLAYER* and *NEXTLAYER* values.

= *LocalTangent* specifies that each location's *NEWLAYER* value should indicate the tangent (multiplied by 100) of the angle represented in degrees by the location's *FIRSTLAYER* value.

= *LocalVariety* specifies that each location's *NEWLAYER* value should indicate the number of dissimilar values associated with that location on *FIRSTLAYER* and any *NEXTLAYER*(s) specified.

The *of FIRSTLAYER* Phrase

of FIRSTLAYER specifies an existing map layer. *FIRSTLAYER* is this layer's title or a numeral. As a numeral, it represents a layer on which all locations are set to that numeral's value.

The *and NEXTLAYER* Phrase

and NEXTLAYER specifies an additional existing map layer. *NEXTLAYER* is this layer's title or a numeral. As a numeral, it represents a layer on which all locations are set to that numeral's value.

A-4 *ZonalFUNCTION* **OPERATIONS**

The *ZonalFUNCTION* operations are introduced in Chapter 6. Each generates a new map layer on which every location is set to a value computed as a specified function of those values from one existing map layer that are associated with all locations within a common zone on another existing layer. *ZonalFUNCTION* statements are specified as

NEWLAYER   = ZonalFUNCTION of FIRSTLAYER [within SECONDLAYER]

The *NEWLAYER* **Phrase**

NEWLAYER is the title to be assigned to the new map layer. If *NEWLAYER* matches the title of an existing map layer, the existing layer is deleted when the new layer is created. The resolution and orientation of *NEWLAYER* are those of *FIRSTLAYER*.

The *ZonalFUNCTION* **Phrase**

= *ZonalFUNCTION* specifies the function to be used in calculating new values. *FUNCTION* is one of the terms *Combination, Majority, Maximum, Mean, Minimum, Minority, Percentage , Percentile, Product , Ranking, Rating, Sum,* or *Variety*.

= *ZonalCombination* specifies that each location's *NEWLAYER* value should indicate the particular combination of *FIRSTLAYER* values that occur at one or more locations within its *SECONDLAYER* zone. Values of one, two, three, and so on are assigned according to the order in which combinations occur when sorted by minimum value.

= *ZonalMajority* specifies that each location's *NEWLAYER* value should indicate which *FIRSTLAYER* value occurs most often within its *SECONDLAYER* zone. If two or more such values are found, -0 is assigned.

= *ZonalMaximum* specifies that each location's *NEWLAYER* value should indicate the highest *FIRSTLAYER* value within its *SECONDLAYER* zone.

= *ZonalMean* specifies that each location's *NEWLAYER* value should indicate the average of the *FIRSTLAYER* values at all locations within its *SECONDLAYER* zone.

= *ZonalMinimum* specifies that each location's *NEWLAYER* value should indicate the lowest *FIRSTLAYER* value within its *SECONDLAYER* zone.

= *ZonalMinority* specifies that each location's NEWLAYER value should indicate which FIRSTLAYER value occurs least often within its SECONDLAYER zone. If two or more such values are found, -0 is assigned.

= *ZonalPercentage* specifies that each location's NEWLAYER value should indicate the number of locations within its SECONDLAYER zone that have equal FIRSTLAYER values. This number is expressed as a percentage of the total number of locations in that zone.

= *ZonalPercentile* specifies that each location's NEWLAYER value should indicate the number of locations within its SECONDLAYER zone that have lower FIRSTLAYER values. This number is expressed as a percentage of the total number of locations in that zone.

= *ZonalProduct* specifies that each location's NEWLAYER value should indicate the multiplicative product of the FIRSTLAYER values at all locations within its SECONDLAYER zone.

= *ZonalRanking* specifies that each location's NEWLAYER value should indicate the number of zones occurring at one or more locations within its SECONDLAYER zone that have lower FIRSTLAYER values.

= *ZonalRating* specifies that each location's NEWLAYER value should be explicitly assigned according to the combination of FIRSTLAYER values occurring within its SECONDLAYER zone. New values are assigned through one or more phrases given as *with NEWVALUE for FIRSTZONES* where

- *NEWVALUE* is a numeral or the title of an existing layer, and
- *FIRSTZONES* is one or more numerals representing FIRSTLAYER values.

As a numeral, NEWVALUE specifies a constant value to be applied to any SECONDLAYER zone that contains (only) the specified FIRSTLAYER value(s). As a title, NEWVALUE indicates that each location's new value is to be set to that of the same location on the map layer whose title is specified. If no new value is specified for a particular SECONDLAYER zone, its locations are set to -0.

= *ZonalSum* specifies that each location's NEWLAYER value should indicate the sum of the FIRSTLAYER values at all locations within its SECONDLAYER zone.

= *ZonalVariety* specifies that each location's NEWLAYER value should indicate the number of FIRSTLAYER zones that occur at one or more locations within its SECONDLAYER zone.

The *of FIRSTLAYER* **Phrase**

of FIRSTLAYER specifies the existing map layer whose values are to be summarized within each *SECONDLAYER* zone. *FIRSTLAYER* is this layer's title or a numeral. As a numeral, it represents a layer on which all locations are set to that numeral's value.

The *within SECONDLAYER* **Phrase**

within SECONDLAYER specifies an existing map layer defining the zones within which *FIRSTLAYER* values are to be summarized. *SECONDLAYER* is this layer's title. If no *SECONDLAYER* is specified, the entire study area is treated as a single *SECONDLAYER* zone.

INDEX

Absolute position, 40, 169-170
Absolute value, 70
Accessibility (*see* Proximity)
Accumulation of travel costs, 140, 142-145, 170-171
Accuracy, 30, 38, 55
Addition (*see* *FocalSum* operations, *LocalSum* operations, *Zonal-Sum* operations)
Algebra, 53, 56, 68, 86, 199, 200
Algorithmic allocation techniques, 208, 212
All-inclusive subcomponents, 192
Allocation, 198-200:
  evaluation of problem solutions, 201, 204, 205, 222-223
  solution of problems, 201, 203-204, 205, 206-222, 224, 225
  statement of problems, 201-202, 205-206
  of individual zones, 201-202, 220-223
  of multiple zones, 201-202, 204, 220
  with individual criteria, 201-202, 205, 207, 210, 220-223
  with multiple criteria, 201-204, 220
Analysis, 168, 169, 177, 190, 194, 196, 198
and NEXTLAYER specifications, 238
Arc Cosine (*see* *LocalArcCosine* operations)
Arc Sine (*see* *LocalArcSine* operations)
Arc Tangent (*see* *LocalArcTangent* operations)
Area, 30-32, 33 (*see also Incremental Area* operations)
Areal characteristics (*see* Areal form)
Areal form, 25, 30-32, 36, 37, 38, 104, 106-107, 116, 177, 178-182

Areal intrusion, 182
Areal protrusion, 182, 183
Areal roundness, 178, 179
Areal smoothing, 119, 121
Aronoff, Stanley, 226
Aspect (*see* Surficial aspect)
at DISTANCE specifications, 118-125, 231
Atomistic, 200, 201-209, 210, 212, 220, 221, 222, 223
Atypical value, functions of, 95, 98, 116, 195

Bearing, 26, 95, 124, 126-128 (*see also FocalBearing* operations)
Berry, Brian J.L., 227
Berry, Joseph K., xii
Best-fitting planes, 108-110, 117, 151
Bird's eye perspective, 50-51, 54, 55
Block encoding, 40
Boolean algebra (*see* Logical functions)
Brown's Pond cartographic model, 4-5, 6, 8, 10, 12-13, 17-21, 40
Brown's Pond study area, 4-5
Buffer generation, 126
Burrough, Peter A., 226
by DIRECTION specifications, 118-125, 232

Cartographic modeling, xi, xiii-xviii, 54, 226, 227
Cartographic models, xi, 3, 4-6, 38, 54, 61
Cartography, xii, 3, 6, 12, 44, 54, 227
Categories, 10
Cells, 17

GIS AND CARTOGRAPHIC MODELING

Mode (*see FocalMajority*
operations, *LocalMajority*
operations, *ZonalMajority*
operations)
Modeling (*see* Cartographic
modeling)
Models (*see* Cartographic models)
Modifiers of statements, 57
Modular difference, 70
Monmonier, Mark S., 227
Multiplication (*see FocalProduct*
operations, *LocalProduct*
operations, *ZonalProduct*
operations)
Mutually exclusive subcomponents,
192

Narrowness, 182, 184-185
Nearest neighbors (see
FocalNeighbor operations)
Neighborhoods, 22, 23, 39
Neighbors (*see* Neighborhoods)
Networks, 140-141, 143-144, 146-148
NEWLAYER specifications, 58-59,
in *FocalFUNCTION* operations,
229
in *LocalFUNCTION* operations,
234
in *IncrementalFUNCTION*
operations, 236
in *ZonalFUNCTION* operations,
239
New map layers, 49, 52
Nodes, network, 140
Noise, 150
Nominal-scale measurements, 14,
15, 45, 73, 76, 78, 84, 95, 98,
102, 116, 195
NonEuclidean, 136, 142, 143 (*see
also* Euclidean)
Normalization, 163
Null values, 12, 14, 16, 56, 84
Numerals, 58

Objectives, 198, 200, 203, 208
Objects of statements, 57
Observations, 17
of FIRSTLAYER specifications, 58-59,
in *FocalFUNCTION* operations,
152, 231
in *LocalFUNCTION* operations,
238

in *IncrementalFUNCTION*
operations, 235
in *ZonalFUNCTION* operations,
241
One-dimensional (*see* Punctual
form)
on SURFACELAYER specifications:
in *FocalFUNCTION radiating*
operations, 132-134, 135,
233
in *FocalFUNCTION spreading*
operations, 146, 147, 232
in *IncrementalFUNCTION*
operations, 107, 235
Optimization, 200, 216, 222
Ordinal-scale measurements, 6, 8,
14, 15, 45, 76, 78, 84, 95, 98,
102, 116, 195
Orientation, 3, 6, 8-9, 24, 36, 107, 108,
109, 197
Outlining, 193
Over-constrained allocation
problems, 203
Overlaying, 72
Overlay mapping, xiii, 226
Overlays, xiii, 6

Percentile, 92 (*see also FocalPer-
centile* operations, *Zonal-
Percentile* operations)
Perimeter, 30-32 (*see also Incremen-
talFrontage* operations)
Petersham, 4
Picture elements, 17
Picture functions, 6
Pixels, 17
Points, 16
Position (see Absolute position,
Relative position)
Precision, 13-14, 36-38
Product (*see FocalProduct*
operations, *LocalProduct*
operations, *ZonalProduct*
operations)
Programming, 44, 46-48, 70
Projection, 24-25
Protrusion (see Areal protrusion,
Surficial protrusion)
Proximal zones, 128
Proximity, 27, 126 (*see also
FocalProximity*
operations, Line-of-sight
proximity, Travel-cost
proximity)